THE THEORY OF
ELECTRICAL CONDUCTION AND BREAKDOWN IN SOLID DIELECTRICS

BY

J. J. O'DWYER

CLARENDON PRESS · OXFORD
1973

Oxford University Press, Ely House, London W.1

GLASGOW NEW YORK TORONTO MELBOURNE WELLINGTON
CAPE TOWN IBADAN NAIROBI DAR ES SALAAM LUSAKA ADDIS ABABA
DELHI BOMBAY CALCUTTA MADRAS KARACHI LAHORE DACCA
KUALA LUMPUR SINGAPORE HONG KONG TOKYO

PRINTED IN NORTHERN IRELAND AT THE UNIVERSITIES PRESS, BELFAST

PREFACE

THIS book is more in the nature of a new work than a second edition of an older one; it is true that much of the material from the original book has been included, but the presentation has been completely rearranged and modified by the inclusion of a great deal of new material. In addition, the scope has been extended to include electrical conduction (particularly high-field conduction) as well as breakdown.

The arrangement reflects the purpose of the book, which is to set out the theories of high-field conduction and dielectric breakdown of solids; experimental work is therefore not presented in any systematic manner, but rather by way of illustration of applications of the theory. Low-field conduction is reviewed briefly (extensive coverage of the topics in this chapter is available elsewhere), and the various theories of high-field conduction are classified and presented more fully. The division of breakdown theories into 'thermal', 'intrinsic', and 'avalanche' has been dropped in favour of the earlier classification 'thermal' and 'purely electrical'; the reasons for taking this step are explained in the text.

In addition to the various examples of experimental work cited throughout, the chapter on special substances gives a brief review of work on alkali halides, the oxides of silicon, and polymers. This selection is highly arbitrary, but is made on the grounds that the first group are the academically interesting dielectrics while the others are leading examples of dielectrics of great practical importance.

I wish to express my thanks to Professor H. Fröhlich F.R.S., who initiated my interest in this subject and suggested the writing of this book.

Acknowledgement is made to the Editors of Advances in Physics, Australian Journal of Physics, Journal of Applied Physics, Journal of Applied Physics Letters, Journal of the Institute of Electrical Engineers, Journal of Non-Crystalline Solids, Journal of Physics and Chemistry of Solids, Journal of Vacuum Science and Technology, Philosophical Magazine,

Physica, Physical Review, Physical Review Letters, Physica
Status Solidi, The M.I.T. Laboratory for Insulation Research,
Proceedings of the Institute of Electrical and Electronics
Engineers, Proceedings of the Royal Society, Progress in
Dielectrics, Solid State Communications, Soviet Physics
Solid State, Thin Solid Films, Transactions of the Metallurgical
Society of the American Institute of Mechanical Engineers,
and Zeitschrift für Physik for permission to reproduce dia-
grams and figures.

J. J. O'D.

Oswego, New York
June 1972

CONTENTS

1

A SURVEY

1.1. Introduction

ALL materials conduct electricity to a greater or lesser extent, and all suffer some form of breakdown in a sufficiently strong electric field. For low field strengths the conduction process in most materials is ohmic, but as the field strength is increased the conductivity usually becomes field-dependent; if the field strength is increased further some form of destructive irreversible conduction takes place. Gases undergo spark or glow discharges depending mainly on the pressure, and liquids suffer at least a temporary loss of their insulating properties; because of the nature of the materials, no permanent discharge track remains in either of these cases. On the other hand, dielectric solids usually exhibit a permanently damaged discharge track, and this occurs whether the breakdown has been due to thermal instability or to other causes. The other causes that may determine dielectric breakdown of solids are still not completely known, and it is one of the objectives of this book to elucidate them as far as possible.

Thermal breakdown was originally identified on the basis of both experimental and theoretical evidence; the type of breakdown that could not be so identified was designated as purely electrical. In the development of the subject purely electrical breakdown was subclassified into intrinsic and avalanche breakdown, again on the basis of combined experimental and theoretical results. More recent developments, however, have been in the direction of combining the notions of intrinsic and avalanche breakdown, thus retrieving the older concept of purely electrical breakdown.

1.2. Electrical conductivity of dielectrics

The conductivity of dielectrics may be either ionic or electronic or both. It may be a matter of great difficulty to separate these components experimentally, particularly in measurements at high field strength; however, the basic theoretical ideas are quite distinct and can easily be treated as such.

Ionic conductivity is simply due to the migration of positive or negative ions. The basic theoretical expression for all electrical conductivity is

$$\sigma = \sum_i n_i e_i \mu_i, \qquad (1.1)$$

where n_i is the density of carriers of the ith species, and e_i and μ_i the corresponding charges and mobilities. In the usual analysis of experimental low-field ionic conductivity, one writes

$$\sigma = \sigma_0 \exp(-\phi/k_0 T), \qquad (1.2)$$

where σ_0 and ϕ are experimentally determined (usually constants within some given range of temperature), k_0 is Boltzmann's constant, and T the absolute temperature. Since both the density and the mobility of the migrating species are usually temperature-dependent, the single temperature-dependent term in eqn (1.2) combines the effects of temperature on both n_i and μ_i of eqn (1.1); this point will be taken up in detail in later chapters. The field-dependence of ionic mobility should be small, but the concentration of charge carriers could be strongly influenced by an electric field. This can arise, for example, if an electric field lowers the activation energy for separation of the constituents of a bound ionic complex; it would be represented empirically by the introduction of field-dependent terms in eqn (1.2).

The simplest understanding of electronic conduction in a solid dielectric arises from modifications to the quantum-mechanical band theory of solids. This gives a picture of a dielectric in which a series of allowed electronic energy bands are completely occupied by electrons up to a certain level, and empty thereafter. The conduction band does not give rise to any conductivity since it contains no electrons, and neither does the valence band conduct since there are no unoccupied states into which an electron can be accelerated by an applied field. This simple picture applies only to a perfect crystalline insulator at the absolute zero of temperature. Real insulators at finite temperature differ from it in very important respects. In the first place isolated electronic energy levels, i.e. electron traps, will be present in what is normally the forbidden band. These trapping levels are caused by the influence of foreign ions, vacant lattice positions, interstitials, etc. on the normal lattice field. Secondly, at any finite temperature, electrons will be thermally excited into conduction levels and traps that would not be occupied at the absolute zero. Clearly the density of electron traps will be determined by the density of deformations of the lattice field that give rise to an effective potential well. In general, only the conduction and valence levels are important in pure, strain-free crystals at low temperatures. The relative importance of isolated levels increases with admixture of foreign ions, mechanical strain, and rise in temperature.

The understanding of electronic conduction in terms of conduction bands and traps is not the only way to proceed; indeed in certain circumstances it is inappropriate. In band theory the effect of an applied electric field is regarded as causing an acceleration of conduction electrons in the band; it may however be more correct to regard the electric field as causing transitions between adjacent but more or less localized states of the conducting electrons. This mechanism is referred to as hopping conduction, and the conducting electron is regarded as making its way through the dielectric in a series of discrete movements. Band conduction and hopping conduction are not regarded as the antithesis of each other; rather the latter is a more appropriate visualization than the former in the limiting case of narrow bands.

These simple ideas of the electronic conductivity of dielectrics fail in a striking way for certain of the transition-metal oxides. Some of these substances have only partly filled bands (and hence should be metallic in the simple picture), yet they are highly insulating; others exhibit metallic conductivity at high temperatures, but are insulating at low temperatures.

Regardless of the theoretical model which one may adopt to explain electrical conduction in the insulating transition-metal oxides, the experimental data is adequately described in terms of an activation energy with the conductivity given by eqn (1.2).

The basic theoretical steady-state conduction equation is

$$j = \sigma F, \tag{1.3}$$

where j is the current density, F the field strength, and σ the conductivity defined by eqn (1.1). The experimentally measured quantities are the current I and the voltage V. The simplest procedure to establish a connection between theory and experiment is to write

$$\left. \begin{array}{l} j = I/A \\ F = V/d \end{array} \right\}, \tag{1.4}$$

where A is the cross-sectional area and d the specimen thickness. Since the breakdown path usually extends over only a tiny fraction of the cross-sectional area, the current density in the breakdown channel is much higher than would be found by the application of eqn (1.4). Nevertheless, there does not appear to be any experimental or theoretical work to test the appropriateness of eqn (1.4) for pre-breakdown currents; it is therefore used for lack of anything better. The correct relation between voltage difference and field strength across the dielectric involves consideration of space charge; this is taken into account in many theories discussed in later chapters.

In addition to the theoretical interpretation which should be given to them, measurements of current–voltage characteristics present various practical difficulties, not the least of which is the very long time required to achieve a steady state. Some elementary remarks on this difficulty can be made.

It is well known (cf. Panofsky and Phillips 1962) that the relaxation time associated with the approach of a current distribution to the steady state is given by

$$\tau = \epsilon/4\pi\sigma \tag{1.5}$$

where ϵ is the dielectric constant and σ the conductivity. This result arises from a consideration of an infinite homogeneous dielectric, so that the conductivity in question has the same connotation. If the current is due to charge carriers of different species the conductivity will be given by eqn (1.1). In this case there will be a relaxation time

$$\tau_i = \epsilon/4\pi n_i e_i \mu_i \tag{1.5a}$$

associated with each species of charge carrier, and the eventual attainment of a steady state will be determined by the longest of the time constants (1.5a). It is readily verified that small concentrations of relatively immobile species may result in relaxation times of order of hours. This will not be important if the space-charge density due to the carriers is so low as to cause negligible field distortion. To estimate this effect let us take a very low mobility of 10^{-10} cm² V⁻¹ sec⁻¹, a dielectric constant of 6, and a relaxation time of 1 hour; from eqn (1.5) this leads to a carrier concentration of 10^{13} cm⁻³. Poisson's equation for the effect of the space charge of the ith species is

$$\partial F/\partial x = 4\pi n_i e_i, \tag{1.6}$$

which leads to a value of 10^7 V cm⁻² for the rate of variation of the field. This could easily lead to observable effects.

However, another reason for long-time variation of the conduction process arises from possible carrier injection from the electrodes. For example electrons may be injected from a blocking contact at the cathode;† these electrons may be trapped as they cross the dielectric towards the anode, and this would cause a space charge tending to reduce injection from the contact. Gradual trapping of electrons at deeper levels could be associated with very long time constants, and the final steady state would depend in a complex way on the details of these processes. Injection of this type results in an electrical modification of the properties of the test sample—in simple language the sample becomes charged. An even more complex

† This process will be discussed at length in Chapter 3.

situation can arise if ionic species are injected from an electrode. An interesting example of this is given by the work of Harris (1968) who decorated dislocations and subgrain boundaries of KBr with silver, after prolonged application of a strong electric field using silver electrodes; the apparent conductivity of the KBr varied continuously over many weeks. It is clear that both an electrical and a compositional change took place in the test specimen during the period of the experiment. For the purpose of this book such decoration of imperfections with foreign ions or atoms is regarded as a hazard to be avoided experimentally, rather than a phenomenon to be explained.

Some general observations can be made about the electrical conductivity of dielectrics, which bear on the mechanism of breakdown. At low temperatures the conductivity is always small, the electronic part for lack of charge carriers and the ionic part by reason of the low mobility of the ions (if not also for other causes). For low field strengths the ionic conductivity (if it exists) is usually greater than the corresponding electronic conductivity, and this will be true for all temperatures. At low temperatures the ionic conductivity is too small to cause more than infinitesimal heating of a dielectric solid, which will then withstand applied fields so strong that electronic conductivity becomes dominant. At high temperatures the enhanced ionic conductivity plays a decisive role in the failure of a dielectric before the field is sufficiently strong to cause appreciable enhancement of the electronic component.

1.3. Breakdown of dielectrics

The early study of dielectric breakdown strength was carried out under conditions in which failure of the dielectric was more often than not caused by discharges in the ambient medium, microscopic flaws in the dielectric itself, or discharges due to intense local field concentration. Regarding these effects as spurious, experimental workers refined techniques so as to eliminate them and measure a breakdown strength genuinely characteristic of the dielectric material and the manner of application of the voltage. Various more or less standard types of test specimen have been developed, and some of the principal ones are described below.

(1) A dielectric specimen in the form of a thin plate of material (often a thin crystal of an alkali halide) is carefully prepared to avoid macroscopic defects and has a spherical indentation made in one face. Evaporated metallic films ensure good contact with a sphere–plane electrode system, and a high-strength ambient such as transformer oil circumvents the difficulty of discharges in the external medium. A typical arrangement of this type is shown in Fig. 1.1.

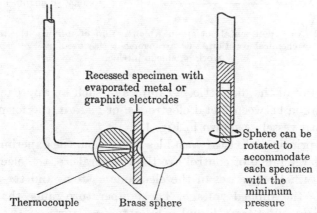

Recessed specimen with evaporated metal or graphite electrodes

Sphere can be rotated to accommodate each specimen with the minimum pressure

Thermocouple   Brass sphere

FIG. 1.1. A typical test specimen of the recessed type used by Cooper (1962) in work on alkali halides.

(2) A thin dielectric film is deposited by some suitable technique (such as vacuum evaporation, oxidation of a surface, etc.) and the electrodes are placed as parallel plates on the opposite surfaces of the film. In most cases at least one of them is an evaporated metal film, while the other is frequently a solid metallic base on which the test specimen was prepared. An experimental arrangement in which the test film was deposited by chemical reaction is shown in Fig. 1.2.

(3) A molten dielectric is introduced between suitable electrodes and allowed to solidify. In some cases such a technique has proved successful, but it clearly depends on special

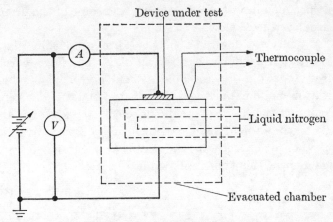

FIG. 1.2. An experimental set-up in which a film of uniform thickness is formed by chemical reaction; this arrangement was used by Sze (1967) in work on silicon nitride.

properties of the dielectric. Examples of such an approach are solid argon between metal electrodes, or glass in the form of a spherical vessel as shown in Fig. 1.3.

The principal physical variables over which the experimenter has a large range of control are the temperature, the electrical and thermal properties of the electrode system, and the wave form of the applied voltage. The temperature range that has been covered is from liquid-helium temperatures up to some hundreds of degrees Celsius; the most satisfactory ambient media are in general the chemically inert liquids and gases, or in certain geometrical configurations a continuation of the dielectric substance under test. The electrical and thermal properties of the electrodes depend on the material used, its manner of application, and its thermal capacity and conductivity. Note for example that the breakdown properties of a thin dielectric film between thin evaporated-metal electrodes may be quite different from the breakdown properties of the same film if one of the electrodes is a relatively massive one, since the heat capacity of the latter electrode arrangement is much greater. The method used in applying the field to the specimen also needs careful specification. The magnitude of an

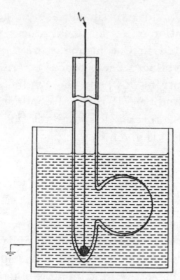

Fig. 1.3. A test specimen of glass in the form of a thin spherical shell; the electrodes were the electrolytes both inside and outside the spherical surface (after Vermeer 1959).

applied d.c. or a.c. voltage may be slowly raised up to break-down—the so called d.c. and steady-state a.c. tests; a series of impulse voltages of slowly rising amplitude (sometimes of alternating polarities) may be applied till breakdown occurs; or a single shot of voltage, rising with time in some given manner, may be used to bring about the breakdown.

In a typical experiment one measures the current–voltage response of the specimen up to the final destructive breakdown; we propose to divide the discussion into two parts depending on whether or not the breakdown can be understood in terms of the pre-breakdown conduction process without the necessity to postulate the occurrence of further physical phenomena in the dielectric such as, for example, collision ionization.

(a) *Breakdown as a continuation of the conduction process— thermal breakdown*

In a sense all breakdown is a continuation of the conduction process, but, in the type of breakdown that has become known as thermal, theory and experiment are in good agreement without the postulation of any physical processes additional to those which operated in a continuous way from the initial application of the voltage. These processes are the Joule heat generated by the current flow, and the conduction of this heat away to the surroundings. If thermal conduction is the only significant heat-loss process, the point relation describing the lattice energy balance is given by

$$C_V \frac{\mathrm{d}T}{\mathrm{d}t} - \mathrm{div}(\kappa \, \mathrm{grad} \, T) = \sigma F^2, \qquad (1.7)$$

where C_V is the specific heat per unit volume, κ is the thermal conductivity, σ is the electrical conductivity, F the field, $\mathrm{d}T/\mathrm{d}t$ is the time derivative, and grad T the space gradient of the temperature. This is the fundamental equation of thermal breakdown; other discussions can be considered as arising from approximations to it. Since σ and κ are always temperature-dependent (the former usually strongly so), and in addition σ may depend on the field strength, even approximate analytic solutions of eqn (1.7) are not possible for any but the simplest boundary conditions. A complete solution would give T as a function of time and position, but all of this information is not required. Since failure of the dielectric will depend on the temperature of its hottest part, a numerical solution need calculate only the temperature of the hottest part as a function of time for some specified manner of application of the field. Numerical solutions of eqn (1.7) for the d.c. case (constant voltage applied at time $t = 0$) show certain general features, which are illustrated in Fig. 1.4. The principal result is that there exists a critical field strength F_m for which the temperature of the hottest part of the dielectric asymptotically approaches some temperature T_m (not necessarily the melting point of the dielectric—in fact often a very much lower temperature) with time. For field strengths greater than F_m the temperature reaches the value T_m in a finite time and thereafter increases without limit, while for lower field strengths the

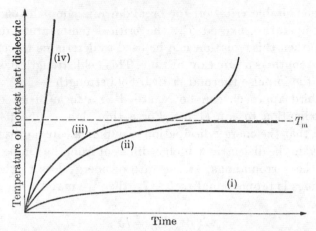

FIG. 1.4. Schematic diagram of solutions to equation (1.7). (i) No breakdown. (ii) Thermal breakdown (1.7a). (iii) Thermal breakdown (1.7). (iv) Thermal breakdown (1.7b).

temperature rises slowly to some upper limit that depends on the field strength.

It is evident that the thermal critical field strength depends on the time application of the field. Various limiting and approximate cases arise that can be treated simply. The first case is that in which there is almost a steady state for the lattice processes with the temperature of the hottest part of the dielectric equal to T_m. The time-dependent term of eqn (1.7) then vanishes and the solution of

$$-\mathrm{div}(\kappa\,\mathrm{grad}\,T) = \sigma F^2 \tag{1.7a}$$

gives the minimum thermal critical field F_m appropriate to the case in which the field is applied for a very long time. In the literature, F_m has simply been called the thermal breakdown strength.

The second case occurs when the field is applied as a short pulse (of the order of seconds duration or less); it may then be a satisfactory approximation to ignore the heat-conduction term in eqn (1.7) and obtain the equation

$$C_V\,\frac{\mathrm{d}T}{\mathrm{d}t} = \sigma F^2. \tag{1.7b}$$

If some suitable criterion for breakdown is adopted (e.g. that the temperature exceed T_m, the critical temperature defined above) then this equation can be used to determine the critical field strength as a function of time. This field strength has been called the impulse thermal critical field strength.

A third approach can be regarded as a form of integrated approximation to eqn (1.7a). For a thin slab or film it is assumed that the energy dissipation results in a constant temperature T in the dielectric, which is different from the temperature T_0 of the surroundings. If the rate of energy loss to the surroundings is proportional to $T-T_0$, then we may write

$$\lambda(T-T_0) = IV, \qquad (1.7c)$$

where I is the current, V the applied voltage, and λ a constant for the particular test set up. This equation is the basis for one approximate approach to the thermal breakdown of thin films; the topic will be discussed at greater length in a later chapter.

The equations proposed above are the chief ones governing thermal breakdown as it has arisen in controlled experimental situations; they do not represent all conceivable cases. It should also be mentioned that in many cases the field will not be uniform within the dielectric, so that it is more logical to speak of a critical voltage than a critical field strength. However, experimental results are frequently expressed in terms of the field strength; this should cause no confusion provided it is understood as a mean field strength, and that non-uniformity of the field may be an important factor in the situation.

Finally a word can be said about the general experimental characteristics of thermal breakdown. They are as follows.

(1) High temperatures favour thermal breakdown, since in general the electrical conductivity increases and the thermal conductivity decreases as the temperature increases.

(2) The thermal breakdown strength depends on the size and shape of the sample, and on the geometry and thermal properties of the electrodes and the ambient medium. It takes milliseconds or longer for breakdown to develop.

(3) In an impulse thermal breakdown, the strength does not depend greatly on the size and shape of the sample, provided the electrode arrangement is such that heat is not conducted away too rapidly. The breakdown strength varies greatly with time of application of the field, being larger for voltage pulses of short duration.

(4) For alternating electric fields the breakdown strength will usually be lower than the d.c. breakdown strength, since the power loss in a dielectric usually increases with frequency.

(b) Breakdown as caused by the onset of collision ionization—purely electrical breakdown

In this section we revert to the older terminology of purely electrical breakdown, since the theories of intrinsic and avalanche breakdown which have been current for some years both have major deficiencies; it seems in fact that some combination of the two offers the best explanation of this type of breakdown. Let us note at the outset that there does not appear to be any direct and convincing evidence that collision ionization is in fact a precursor of dielectric breakdown; however, it does seem to be reasonable to assume that it is so, since, in the first place, collision ionization is known to occur in semiconductors, which differ from dielectrics chiefly by having a narrower forbidden band gap, and, in addition, it is difficult to visualize an alternative triggering mechanism which fits in naturally with the known properties of solids. From this point of view the theories of intrinsic breakdown yield values of the field strength for which collision ionization becomes an important process, and the theories of avalanche breakdown describe how the products of the collision ionization (i.e. the electrons and holes) cause the build-up of breakdown current.

The intrinsic critical field strength is conceived as being the field strength for which some instability occurs in the electronic conduction current. One considers a uniform field acting on a dielectric of unspecified extent, the influence of the electrodes being completely ignored.

The first such calculation of an intrinsic critical field was given by Zener (1934), who calculated the rate of quantum-mechanical tunnelling from the valence band to the conduction band in the presence of a strong electric field. The Zener tunnelling current is a strongly increasing function of the field strength, and the instability criterion was chosen arbitrarily as the value of the field for which the current from the valence band exceeded some assigned value.

More detailed calculations of the intrinsic critical field strength have considered the conduction-electron energy-balance equation. If an electric field F is applied to a dielectric, and causes a current flow of density j, then the rate of energy gain from the field is

$$A = jF. \tag{1.8}$$

If some mechanism exists whereby conduction electrons can transfer energy to the lattice, we call this rate of energy transfer B. The rate A depends on F and on the lattice temperature T, while the rate B depends on T. In addition, both A and B depend on parameters describing the conduction electrons. If we denote these parameters collectively by α, the condition for energy balance in the field is

$$A(F, T, \alpha) = B(T, \alpha), \tag{1.9}$$

in so far as the conduction electrons are concerned. Since thermal processes are being neglected, the slight temperature rise of the lattice consequent on Joule heating is ignored and T treated as a constant in eqn (1.9). Depending on details of the model, it may then occur that eqn (1.9) can be satisfied in a physically acceptable manner only for values of F below a certain critical value F_c. This value is regarded as the intrinsic critical field for the particular model being considered (cf. von Hippel 1935, Fröhlich 1937, 1947a, Callen 1949 and Fröhlich and Paranjape 1956). The models proposed by these authors differ from each other by considering different mechanisms of energy transfer from the conduction electrons to the lattice, and also by the different assumptions they make concerning the energy distribution of the conduction electrons. In an inelastic collision with the lattice, a conduction electron may be scattered into another conduction state, or trapped at a localized energy level. As far as the description of conduction electrons is concerned, calculations of intrinsic breakdown must always consider a single-electron approximation or a hot-electron theory. The reason for this is that any instability exhibited by a conduction-electron distribution whose temperature is the same as the lattice temperature is thermal breakdown.

The intrinsic critical field strength therefore marks a more or less abrupt transition from circumstances in which collision ionization is negligible to those in which it is not; an important feature of calculations of intrinsic critical fields is that they should yield the temperature dependence of the field strength which marks this large increase in collision ionization.

Theories of avalanche breakdown arose as an attempt to take explicit account of the existence of the electrodes as the boundaries between which the electric discharge propagates. The simplest form of avalanche breakdown theory due to Seitz (1949) considers the conditions in which a single electron (or very few electrons) starting at the cathode can cause an avalanche of electrons of sufficient size to destroy the dielectric. If a single electron leaving the cathode can succeed in producing another conduction electron by collision ionization, and these two produce a further two, an avalanche of n electrons will be produced in n generations. If the critical sized avalanche can be estimated (from thermal considerations), then a knowledge of the mean free path for collision ionization (obtained from considerations such as those employed in calculating intrinsic critical fields) would give the inter-electrode distance required for such an avalanche to build up. This type of theory of avalanche breakdown attempts to incorporate the more soundly based features of the intrinsic and thermal theories into one, at least for those cases in which the initial instability is electronic in nature.

Another main line of development of avalanche breakdown theory has resulted from the realization that space charges (both electrons and holes) caused by the build-up of an electron avalanche should result in a non-uniform field-strength distribution. If one drops the assumption of a uniform field (which was made in the early theories) and treats the hole current as well as the electron current, then a natural basic theoretical assumption is continuity of current. There have been several attempts to formulate such a theory (cf. O'Dwyer 1967, 1969a) and all have a common feature; they envisage breakdown as occurring when the space charge immediately in front of the cathode (due to relatively immobile holes) reaches such proportions that the electron current injected from the cathode is sufficient to destroy the material.

Both types of avalanche theory have their weaknesses; as discussed above, the basic assumptions have been uniformity of the field strength or continuity of the current. The first assumption is self-contradictory in the presence of collision ionization, the second can be made consistent, but the truth lies probably somewhere in the middle. For thin films there would normally be few generations of collision ionization and the uniform-field assumption should be approximately correct provided that the current is not too large. Distortion of the field depends on the charge density, and even a single collision ionization between cathode and anode would make the distortion large if the current were large. On the other hand, it is unlikely that field distortion can ever be ignored in thicker films, in which relatively many generations of collision ionization will occur; the assumption of continuous current is therefore probably a better starting point, although not an adequate one.

The general experimental features of purely electrical breakdown, and the circumstances under which it occurs, can be summarized as follows.

(1) It occurs at low temperatures—for many substances this means room temperature or lower.

(2) The breakdown field strength does not appear to depend on the electrode material, if the electrodes are metallic.

(3) To a certain extent, crystalline dielectrics show preferential directions for the formation of the discharge path.

(4) The breakdown field strength is not a function of the voltage wave form from d.c. to single-shot impulses with microsecond rise times or greater; it is therefore inferred that the breakdown process occurs in a time of order of microseconds or less.

(5) The breakdown field strength depends on the thickness of the dielectric; for thick samples the breakdown strength is an extremely slowly varying function of thickness, for thin samples the variation is more rapid.

IONIC AND ELECTRONIC
CONDUCTIVITY AT LOW FIELDS

2.1. Ionic conduction

IONIC conduction in a dielectric is the process in which electric current is carried by the motion of ions. This could arise in a dielectric in two different manners; in an ionic crystal such as an alkali halide the basic constituents are ions, and physical imperfections alone can be responsible for mechanisms of current flow, while in a non-ionic substance chemical imperfection is required to supply the mobile species. The ionic conductivity of ionic crystals has been expounded at length by Lidiard (1957) and will be briefly sketched here, while the ionic conductivity due to impurities in non-ionic solids has received only an empirical treatment. For substances of the latter type, experimental data are fitted to eqn (1.2) to find the activation energy of the conductivity. In the following, we shall therefore concern ourselves solely with conduction in ionic crystals.

The alkali halides are an important and much investigated class of insulating ionic crystal, and we shall arrange the discussion around them. The results of Kelting and Witt (1949) are typical for conductivity as a function of temperature; they are shown in Fig. 2.1 for the case of KCl containing small amounts of $SrCl_2$ and $BaCl_2$. The principal features of these results are a straight-line variation of log σ against $1/T$ for the high-temperature, high-purity case, and an approximate straight line for the low-temperature, highly doped case. The first of these regions is referred to as the intrinsic region since the conduction properties are those of the chemically pure crystal; the second region is called extrinsic since the conductivity depends on the nature and concentration of impurities. It is clear from Fig. 2.1 that the transition temperature between these two regions is a function of the impurity content.

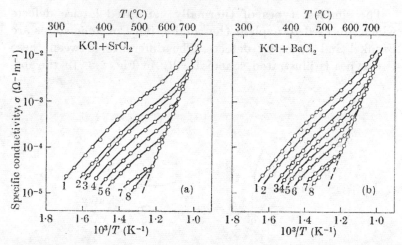

FIG. 2.1. The temperature dependence of the ionic conductivity of KCl containing small amounts of $SrCl_2$ and $BaCl_2$, the quantities as mole fraction (G) being as follows. (a) $G_1 = 19 \times 10^{-5}$, $G_2 = 8.7 \times 10^{-5}$, $G_3 = 6.1 \times 10^{-5}$, $G_4 = 3.5 \times 10^{-5}$, $G_5 = 1.9 \times 10^{-5}$, $G_6 = 1.2 \times 10^{-5}$; (b) $G_1 = 14 \times 10^{-5}$, $G_2 = 9 \times 10^{-5}$, $G_3 = 4.67 \times 10^{-5}$, $G_4 = 3 \times 10^{-5}$, $G_5 = 1.9 \times 10^{-5}$, $G_6 = 1.25 \times 10^{-5}$. Curves 7 and 8 in both cases were obtained for undoped crystals. (After Kelting and Witt 1949.)

An approximate empirical relation for the conductivity can be written (in place of eqn (1.2))

$$\sigma = \sigma_e + \sigma_i$$
$$= \sigma_{0e} \exp(-\phi_e/k_0 T) + \sigma_{0i} \exp(-\phi_i/k_0 T), \qquad (2.1)$$

where ϕ_e and ϕ_i are the conduction activation energies for the extrinsic and intrinsic regions respectively. The constant σ_{0e} depends on impurity content, while σ_{0i} depends only on the host crystal. One would therefore expect the ratio σ_{0e}/σ_{0i} to be much less than unity, and of the order of the impurity fraction; this is borne out experimentally. It is also found from experiment that $\phi_e/\phi_i \sim \frac{1}{2}$; this is explained by noting that ϕ_e should be the activation energy for motion of the extrinsically introduced defect, while ϕ_i contains in addition the activation energy for creation of the defect within the originally perfect crystal.

The simplest types of thermally generated lattice defects which are used to explain the conductivity of ionic solids are Frenkel and Schottky defects. The difference between these two types is illustrated schematically in Fig. 2.2. In the for-

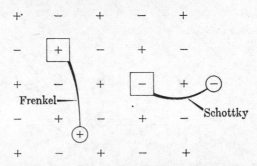

Fig. 2.2. Diagram illustrating the formation of Frenkel and Schottky defects in an ionic crystal. The square indicates that the enclosed ion is missing from the site in question.

mation of a Frenkel defect an ion which was originally at a site of the perfect lattice moves to an interstitial position; the net result of the process is to generate two imperfections—a vacant lattice site and an interstitial ion. A Schottky defect is formed by the migration of an ion which was originally at a site of the perfect lattice to a surface position; the net result of this is to generate only one imperfection—the vacant lattice site. However, if it is required that both the volume and the surface of the crystal remain on the average electrically neutral, then Schottky defects must be created in pairs; the anion and the cation vacancy maintain electrical neutrality in the volume of the crystal, and the displaced anion and cation on the surface. The question as to whether the predominant defects are of Frenkel or Schottky type in a given substance depends on the relative magnitudes of the free energies of formation. It is generally believed that Schottky defects are present in most of the alkali halides at concentrations far in excess of that for Frenkel defects (cf. Lidiard 1957), and we shall assume this to be true for the purpose of the discussion. We shall first give a simple theory of conduction that is adequate to explain the approximate empirical result (2.1), and then a more detailed theory which will be required later for a discussion of the field dependence of the conductivity.

(a) Simple theory of ionic conductivity

As pointed out above there are two tasks involved in the calculation of ionic conductivity; first one must calculate the density of each species of mobile vacancy, and then calculate the appropriate mobility. Equation (1.1) then yields the desired result. We shall perform these tasks in the reverse order, since the same simple approximate theory of mobility is usually applied to vacancy species of all types, while the question of the number densities is more complex.

Consider first a one-dimensional model in which a vacancy migrating in a crystal lattice does so by a series of jumps from one position to the next over a potential barrier; the potential barriers are all the same and the equilibrium positions are all equivalent. It can be shown from classical statistical mechanics (cf. Lidiard 1957) that the probability per unit time for a vacancy to make the transition to a neighbouring equilibrium position is given by

$$w = \nu_0 \exp(-\Delta g/k_0 T), \qquad (2.2)$$

where ν_0 is a frequency which is interpreted as the vibration frequency of the ions surrounding the vacancy, and Δg is the Gibbs free energy of activation. The Gibbs rather than the Helmholtz free energy is used, since it is assumed that the theory relates to experiments at constant pressure rather than at constant volume. The Gibbs free energy can be written in terms of the corresponding enthalpy Δh and entropy ΔS as

$$\Delta g = \Delta h - T\Delta S.$$

The calculation of the jump probability in the absence of a field is carried out with the assumption of thermodynamic equilibrium; in the presence of an applied electric field this assumption can no longer be true, but the resulting steady state can be treated approximately by considering the effect of the electric field as a small perturbation.

Consider the case of a cation vacancy (effective charge $-e$) moving from one equilibrium position to another in the presence of a uniform field F directed along the x-axis (cf. Fig. 2.3);

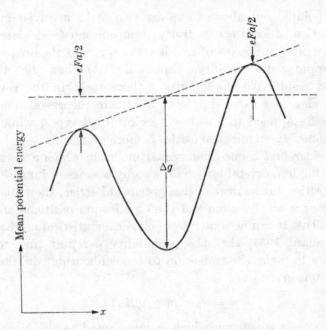

Fig. 2.3. A schematic representation of the mean potential energy of a cation vacancy with an electric field applied in the x direction.

the effect of the field is to add a term eFx to the potential energy. A jump in the direction of the field now takes place with decreased probability

$$w' = \nu_0 \exp\{-(\Delta g + eFa/2)/k_0 T\},$$

and a jump against the field with increased probability

$$w'' = \nu_0 \exp\{-(\Delta g - eFa/2)/k_0 T\}.$$

The mean drift velocity u (in the direction of positive current flow) is therefore given by

$$u = a(w'' - w')$$
$$= a\nu_0 \exp(-\Delta g/k_0 T) \times 2 \sinh(eFa/2k_0 T). \qquad (2.3)$$

Since we are considering the case of low field strength, we assume $eFa \ll k_0 T$; approximating the last term in eqn (2.3), we get

$$u = (a^2 e\nu_0 F/k_0 T)\exp(-\Delta g/k_0 T), \qquad (2.4)$$

which corresponds to a mobility

$$\mu = (a^2 e\nu_0/k_0 T)\exp(-\Delta g/k_0 T). \qquad (2.5)$$

Formula (2.4) holds for a one-dimensional model; for the motion of a cation vacancy in a three-dimensional NaCl-type lattice we refer to Fig. 2.4. Consider the electric field to be

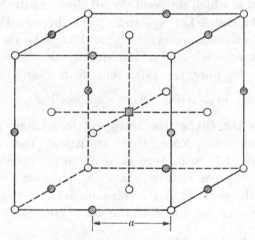

FIG. 2.4. Diagram illustrating the jumps possible for a cation vacancy (shaded square) to nearest-neighbour cation positions (shaded circles) in an NaCl-type lattice. The anion positions are shown as open circles. The distance a is the anion–cation separation.

applied in the (100) direction; the central cation vacancy can jump to any one of the twelve nearest neighbour cation sites, each of which is distant $a\sqrt{2}$ away from the vacancy. The electric field makes no change in the energy requirements for a jump in the transverse direction, but neither do these jumps cause any flow of current. Of the remaining eight possible jumps, four are in the field direction and four are against the field direction; the only change in the algebra is therefore the appearance of a factor 4 in eqns (2.3), (2.4), and (2.5). The last will now read

$$\mu = (4a^2 e v_0/k_0 T)\exp(-\Delta g/k_0 T). \tag{2.6}$$

The same result is true if the electric field is applied in an arbitrary direction, since NaCl-type crystals have isotropic conductivity tensors.

The discussion on the density of mobile species falls naturally into two parts, corresponding to the intrinsic and extrinsic regions of conduction. We treat the intrinsic region first by considering the densities of anion and cation vacancies in a pure crystal in which the condition of electroneutrality requires that they be equal. Let n_{AV} and n_{CV} be the densities of anion and cation vacancies respectively, and let N be the density of possible vacancy sites (i.e. the density of either anions or cations in the pure crystal); then from statistical thermodynamics

$$(n_{AV}/N)(n_{CV}/N) = \exp(-g_S/k_0 T), \tag{2.7}$$

where g_S is the Gibbs free energy of formation of a pair of Schottky defects. Note that statistical thermodynamics requires eqn (2.7) to be true even if the presence of divalent impurities requires that n_{AV} and n_{CV} be not equal. Using results of the form (2.6) for the mobilities of the anion and cation vacancies, we find with the use of eqns (1.1) and (2.7)

$$\begin{aligned}\sigma_i &= n_{AV} e \mu_{AV} + n_{CV} e \mu_{CV} \\ &= \frac{4Na^2 e^2}{k_0 T} \exp(-g_S/k_0 T) \times \\ &\quad \times \{\nu_{AV} \exp(-\Delta g_{AV}/k_0 T) + \nu_{CV} \exp(-\Delta g_{CV}/k_0 T)\}, \end{aligned} \tag{2.8}$$

where the subscripts 'AV' and 'CV' on previously defined quantities refer to those quantities for anion and cation vacancies respectively. Equation (2.8) is still not of the form

$$\sigma_{0i} \exp(-\phi_i/k_0 T)$$

which is the intrinsic part of the empirical conductivity relationship (2.1); however it can be made close in form if the mobility of one vacancy species far exceeds that of the other. Suppose for example that $\Delta g_{CV} \ll \Delta g_{AV}$, so that to a high degree of approximation only the cation vacancies move; then, making the appropriate definitions of Gibbs free energy in terms of enthalpy and entropy

$$g_S = h_S - TS_S$$

and

$$\Delta g_{CV} = \Delta h_{CV} - T\Delta S_{CV},$$

we have

$$\sigma_i = \frac{4Na^2e^2\nu_{CV}}{k_0T} \times$$
$$\times \exp\{(\Delta S_{CV} + S_S/2)/k_0\} \times \exp\{(\Delta h_{CV} + h_S/2)/k_0T\}. \quad (2.9)$$

Some comments should be made on this result.

(1) It is the usual practice to plot experimental results as $\log \sigma$ against $1/T$. However, the pre-exponential factor of eqn (2.9) is inversely proportional to the temperature, so that plots of $\log \sigma T$ against $1/T$ should be made for comparison with the theory.

(2) If both anion and cation vacancies are mobile, then the approximation made in deriving (2.9) from (2.8) is not valid; it is necessary to return to (2.8), and the plot of $\log \sigma T$ against $1/T$ will be a straight line only in the case of equal activation enthalpies for mobility of the anion and cation vacancies.

(3) In comparisons (2.9) with experimental data, it should be borne in mind that possible contributions to the current from all sources except Schottky defects have been ignored.

A selection of experimental values for the various activation enthalpies is shown in Table 2.1 for a few of the alkali halides. In the final column we have listed

$$\phi_i = \Delta h_{CV} + h_S/2,$$

which is the total conduction activation enthalpy due to cation vacancies, and also the quantity $\Delta h_{AV} + h_S/2$, which is the total conduction activation energy due to anion vacancies.

TABLE 2.1

The activation enthalpies for various alkali halides. Some of the figures given refer simply to the fitting of eqn (2.1) to experimental data by inspection; for others a computer programme was used to determine theoretical parameters giving best fit to the data on the basis of different models.

Substance	Formation of Schottky pair	Motion of anion vacancy	Motion of cation vacancy	Conductivity	Source
NaF	2·56	1·54	0·97	2·82 anion 2·25 cation	Bauer and Whitmore (1970)
NaCl	1·96				Etzel and Maurer (1950)
KCl	2·26	1·04	0·71	2·17 anion 1·84 cation	Beaumont and Jacobs (1966)
KCl				2·36 anion 1·66	Fuller, Reilly, Marquardt, and Wells (1968)
KCl			0·77		Gründig (1960)
KBr			0·65		Gründig (1960)
KBr	2·53	0·87	0·66	2·13 anion 1·93 cation	Rolfe (1964)
KI	1·56	1·50	0·72	2·28 anion 1·50 cation	Jain and Parashar (1969)
RbCl				2·55 anion 1·58 cation	Fuller and Reilly (1967)

We turn now to the conductivity in the extrinsic region, which is ascribed to the effect of divalent cationic impurities (cf. Fig. 2.1); the simplest assumption to make is that electro-neutrality is preserved simply by the presence in the crystal of a compensating density of cation vacancies. In eqn (2.7), n_{AV} and n_{CV} are no longer equal but are subject to the condition

$$n_{CV} = n_{AV} + n_I, \tag{2.10}$$

where n_I is the density of divalent cationic impurity. Substituting (2.10) into (2.7) we find

$$\frac{n_{CV}}{N} \frac{(n_{CV} - n_I)}{N} = \exp(-g_S/k_0 T), \tag{2.11}$$

which is a quadratic equation in n_{CV} whose solution is

$$n_{CV} = \frac{n_I}{2} \left[1 + \left\{ 1 + \frac{4N^2 \exp(-g_S/k_0 T)}{n_I^2} \right\}^{\frac{1}{2}} \right]. \tag{2.12}$$

The conductivity is therefore given by

$$\begin{aligned}
\sigma_e &= n_{AV} e \mu_{AV} + n_{CV} e \mu_{CV} \\
&= Ne(\mu_{AV} + \mu_{CV}) \exp(-g_S/2k_0 T) \times \\
&\quad \times \left[\left\{ 1 + \frac{n_I^2}{4N^2 \exp(-g_S/k_0 T)} \right\}^{\frac{1}{2}} + \right. \\
&\quad \left. + \frac{n_I}{2N \exp(-g_S/k_0 T)} \frac{\mu_{CV} - \mu_{AV}}{\mu_{CV} + \mu_{AV}} \right],
\end{aligned} \tag{2.13}$$

from eqns (2.11) and (2.12). Recalling that the intrinsic conductivity (2.8) is equivalent to

$$\sigma_i = Ne(\mu_{AV} + \mu_{CV}) \exp(-g_S/k_0 T),$$

we have from (2.13)

$$\sigma_e = \sigma_i \frac{n_I}{N \exp(-g_S/2k_0 T)} \frac{\mu_{CV}}{\mu_{CV} + \mu_{AV}}. \tag{2.14}$$

In deriving eqn (2.14) we assumed that

$$n_I \gg 2N \exp(-g_S/2k_0 T),$$

or that the impurity density far exceeds the thermally gener-
ated Schottky defect density. If we again assume that $\mu_{CV} \gg \mu_{AV}$ eqn (2.14) reduces to

$$\sigma_e = n_I e \mu_{CV}, \tag{2.15}$$

which is a quite obvious result in view of the approximations.
Using (2.6) in (2.15) we obtain

$$\sigma_e = \frac{4 n_I a^2 e^2 \nu_{CV}}{k_0 T} \exp\left(\frac{\Delta S_{CV}}{k_0}\right) \exp(-\Delta h_{CV}/k_0 T), \tag{2.16}$$

which is approximately of the empirical form

$$\sigma_{0e} \exp(-\phi_e/k_0 T).$$

The activation enthalpy of conduction ϕ_e is then identified
with the values in Table 2.1 giving the activation enthalpies
for the motion of a cation vacancy. A comparison of eqns (2.9)
and (2.16) sufficiently explains the experimentally observed
orders of magnitude of the ratios for σ_{0e}/σ_{0i} and ϕ_e/ϕ_i.

In fact, most of the work on thermal dielectric breakdown
requires nothing more than the approximate empirical expression (2.1), which is obtained by drawing tangents to experimental data such as those shown in Fig. 2.1. The principal
reason for looking more closely at the theory is that field-
dependent conductivity cannot be satisfactorily explained on
this simple basis. We require an even more elaborate description
of the conductivity in the extrinsic region, and now proceed
to this question.

(b) A more detailed theory of ionic conductivity

In order to find a more detailed theory of conductivity in
the extrinsic region, we need to consider interactions between
the various defects; defects of opposite charge may possibly
form associated complexes, and all charged defects will interact
via the long-range Coulomb forces. The interactions are thus
naturally considered in two steps: the equilibrium number of
associated complexes are calculated with the help of the
mass-action formula, and the Coulomb interactions among the

unassociated impurity ions and vacancies are taken into account by a Debye–Hückel type formulation. We shall discuss the first of these topics as being both the simpler and also the more relevant for a later discussion of high-field conductivity; what follows is essentially the treatment of Stasiw and Teltow (1947) as modified by Lidiard (1957).

Equation (2.7) again relates the densities of two intrinsic defects in the unassociated state to the Gibbs free energy of formation of a Schottky pair. As above, let the density of divalent cationic impurity be n_I, and let the concentration of complexes formed by the association of a divalent cationic impurity and a cation vacancy be n_K; then the law of mass action gives

$$\frac{n_K}{n_{CV}(n_I - n_K)} = \frac{12}{N}\exp(\zeta/k_0 T), \qquad (2.17)$$

where ζ is the Gibbs free energy of association of a divalent cationic impurity and a cation vacancy. The factor 12 in eqn (2.17) arises since there are 12 nearest-neighbour positions for the cation vacancy and there are therefore 12 distinguishable orientations of the complex for each sub-lattice position (cf. Fig. 2.4). We have implied in writing down eqn (2.17) that the impurity–vacancy pair are associated if they occupy nearest-neighbour sites on the sub-lattice, and are otherwise dissociated; this is not strictly true but should lead to qualitatively correct results, since a generalization would replace the right-hand side of (2.17) by a sum of terms all of the same form. The equation of electro-neutrality now becomes

$$n_{CV} - n_{AV} = n_I - n_K. \qquad (2.18)$$

We require solutions of eqns (2.7), (2.17), and (2.18) for n_{AV} and n_{CV}; as before the conductivity is then determined from

$$\sigma_e = n_{AV} e \mu_{AV} + n_{CV} e \mu_{CV}. \qquad (2.19)$$

Let us define

$$\xi = (n_{CV}/n_{AV})^{\frac{1}{2}} \qquad (2.20)$$

and

$$H = 12\exp\{(\zeta - g_S/2)/k_0 T\}. \qquad (2.21)$$

The equation determining ξ is then found from (2.7), (2.17), and (2.18) to be cubic,

$$n_I \exp(g_S/2k_0T) = N(\xi - 1/\xi)(1 + H\xi). \qquad (2.22)$$

In terms of the intrinsic conductivity of the pure crystal given by

$$\sigma_i = Ne\{\exp(-g_S/2k_0T)\}(\mu_{AV} + \mu_{CV}), \qquad (2.23)$$

the equation (2.19) becomes after a little algebra

$$\sigma_e = \sigma_i \left(\frac{\xi + \phi/\xi}{1 + \phi}\right), \qquad (2.24)$$

where

$$\phi = \mu_{AV}/\mu_{CV}. \qquad (2.25)$$

So far, the treatment of the model has been exact, and eqn (2.24) has furnished a starting point for computational analysis of careful experimental data (cf. Etzel and Maurer 1950, Beaumont and Jacobs 1966); however we wish ultimately to modify this theory to include the effect of field strength on the conductivity, and this will require further progress analytically using certain approximations. To this end we shall confine ourselves to the case of high impurity density and relatively immobile anion vacancy. Then from eqn (2.20) we have $\xi \gg 1$ and from eqn (2.25) we have $\phi \ll 1$; the use of these inequalities in eqn (2.23) yields

$$\sigma_e = \sigma_i\xi, \qquad (2.26)$$

with ξ now given by a quadratic

$$n_I \exp(g_S/2k_0T) = N\xi(1 + H\xi). \qquad (2.27)$$

In many cases we may have $H \ll 1$ corresponding to

$$(g_S/2 - \zeta) \gg k_0T,$$

but the product $H\xi$ could well be of the order of unity. The degree of association is readily found to be

$$\frac{n_K}{n_I} = \frac{H\xi}{1 + H\xi}, \qquad (2.28)$$

so that $H\xi < 1$ corresponds physically to a small degree of association of divalent impurities and cation vacancies, while

$H\xi > 1$ corresponds to a large degree of association. We treat these two cases separately.

For a small degree of association, $H\xi < 1$ and an iterative solution of the quadratic (2.27) yields

$$\xi = \frac{n_I}{N} \exp(g_S/2k_0 T) \times \left\{1 - 12\frac{n_I}{N}\exp(\zeta/k_0 T)\right\}. \qquad (2.29)$$

Using eqn (2.25), we recover (2.14) as the first approximation when $\mu_{CV} \gg \mu_{AV}$, while the term in curly brackets in (2.29) yields a small correction to this result.

If $H\xi > 1$, the appropriate approximation does not reduce to any of the previous results in limiting cases. Equation (2.27) yields as a first approximation

$$\xi = (n_I/12N)^{\frac{1}{2}}\exp\{(g_S - \zeta)/2k_0 T\}. \qquad (2.30)$$

Using eqns (2.23) and (2.26) (with the approximation $\mu_{CV} \gg \mu_{AV}$), we find
$$\sigma_e = (n_I N/12)^{\frac{1}{2}}e\mu_{CV}\exp(-\zeta/2k_0 T). \qquad (2.31)$$

For the simple result (2.15), the conduction activation enthalpy arises solely from the term in μ_{CV}; eqn (2.31) gives an additional term arising from the free energy of association of the complexes.

There have been various estimates of this free energy of association, and some figures are collected in Table 2.2. The results of Bassani and Fumi (1954) are theoretical calculations based on a simple model of an ionic lattice; the remaining results are derived from conductivity measurements. Those due to Etzel and Maurer (1950), Beaumont and Jacobs (1966), Rolfe (1964), and Gründig (1960) are based essentially on the application of equation (2.24); that due to Lidiard (1957) uses a more elaborate theory including Coulomb interactions between free defects.

2.2. Electronic conduction in general

In the remainder of this chapter we shall discuss electronic conduction in insulators; since the subject is quite complex it will be broken up under various headings, and certain basic assumptions and methods of treatment will be used throughout. The main headings will be polaron conduction, conduction

TABLE 2.2

Energies of association of divalent impurity cations and cation vacancies in various alkali halides.

Substance	Divalent impurity				Source
	Cd^{++}	Ca^{++}	Sr^{++}	Ba^{++}	
NaCl	0·38	0·38	0·45		Bassani and Fumi (1954)
	0·25				Etzel and Maurer (1950)
	0·34				Lidiard (1957)
KCl	0·32	0·32	0·39		Bassani and Fumi (1954)
			0·42		Beaumont and Jacobs (1966)
		0·52			Gründig (1960)
KBr		0·56			Gründig (1960)
KI				0·26	Jain and Parashar (1969)

in non-polar substances, and impurity conduction. In each case the applicability of the band scheme will be discussed first, followed by a treatment of hopping conduction.

(a) Single-electron band conduction

For simplicity we shall consider that the effect of a perfectly periodic lattice field can be accounted for by relating the energy E to the momentum $\hbar k$ using an effective mass m^* so that

$$E = \hbar^2 k^2 / 2m^*, \tag{2.32}$$

which is analogous to the free-electron case. It is well known that in many semiconductors such as germanium and silicon the energy–wave-number relation is not parabolic and the surfaces of constant energy are not spherical (cf. Blatt 1968); nevertheless we persist with eqn (2.32) not only because it is the simplest assumption, but also because detailed knowledge of the energy-band structure is lacking for most dielectrics. It follows then that a conduction electron in a band will have the usual properties of a free electron except that the free-electron mass m will be replaced by the effective mass m^*. In particular, the density of quantum states per unit volume per unit energy interval will be given by

$$g(E) = (2m^*)^{\frac{3}{2}} E^{\frac{1}{2}} / 2\pi^2 \hbar^3. \qquad (2.33)$$

Consider now the effect of deviations from perfect periodicity, such as lattice defects and thermal motion of the lattice; the quantized thermal vibrations of the lattice are called 'phonons', and we shall be concerned chiefly with that part of the conductivity which can be explained in terms of electron–phonon interaction.

The sources of the interaction energy have been discussed by Fröhlich and Seitz (1950), who distinguish three possible contributions to it for the case in which the electron has insufficient energy to cause ionization from the valence band. They are as follows.

(1) Interaction with a dipolar field which arises from a consideration of the lattice vibrations of an ionic crystal as being those of a lattice of point charges. A longitudinal mode of vibration of such a lattice is a longitudinal wave of electric polarization.

(2) Interaction with a dipolar field which arises from the distortion of the electron shells associated with the vibration of the lattice; this interaction is closely related to (1).

(3) Interaction with a short-range non-dipolar component of the field arising from the distortion of the electron shells.
In polar crystals all three interactions occur, while in non-polar crystals there are no dipolar field contributions. Various models of the crystal take into account different members of the three contributions mentioned.

In the case of polar crystals, if the lattice is treated as a set of point charges each having the ionic charge and mass (cf. Fröhlich 1937), the interaction contribution (1) is the only one considered and the distortion of the electron shells is neglected. This neglect of electron-shell distortion is a reasonable approximation if the displacement of neighbouring ions from their equilibrium positions is in the same sense and of about equal amount. These long-wavelength mechanical modes are short-wavelength modes of electrical polarization caused by the alternation of sign of the ionic charges, so that the point-charge approximation is a good one when the most important processes being considered involve the short polarization waves of the lattice. Another model of a polar crystal, due to Fröhlich, Pelzer, and Zienau (1950), considers the vibrations as those of a continuum of known high-frequency and low-frequency dielectric constant and lattice vibration frequency. In this way, interactions (2) and (3) are taken into account in a manner that leads to a good approximation when the most important processes involve polarization waves of long wavelength. In yet another approach, Callen (1949) modified the point-charge approximation by considering the ions to carry effective charges. These effective charges are determined from macroscopic considerations, so that, although the details of the calculation differ, the results are the same as those obtained from the continuum approximation.

For the case of non-polar crystals only the interaction energy (3) is considered; it is determined as in the conduction theory of metals.

The theory of electrical conductivity is essentially the theory of the interaction between these two quantum systems, viz. the electrons and the phonons. If this interaction is weak it can be conveniently treated using the well-known results of perturbation theory; however in many important cases the interaction is not weak and a much more sophisticated theoretical approach is necessary. We shall sketch the perturbation-theory approach, and simply quote conclusions of more complex theories, indicating whether or not they yield identical results.

The general results obtainable from the application of perturbation theory are easily summarized. Consider a single electron interacting with a dielectric medium, which is represented by a set of harmonic oscillators whose angular frequency ω is a function of their wave number w. It should be emphasized that, even though the initial step in this approach to calculating the conductivity is a quantum-mechanical perturbation calculation, the whole theory must be regarded as semiclassical. The reason is that the results of the perturbation theory are applied to find the conductivity in terms of a particle model. The Hamiltonian may be written

$$H = H_{el} + H_{osc} + H_{int}, \tag{2.34}$$

and, provided we confine our attention to transitions in which the electron absorbs or emits only one phonon, the transition probability may be written

$$P_w = \frac{2\pi}{\hbar} |M|^2 \, \delta(\xi), \tag{2.35}$$

where w is the wave vector of the phonon involved in the transition, M is the matrix element of H_{int} between the initial and final states, and

$$\xi = E_{k'} - E_k + (n_{w'} - n_w)\hbar\omega. \tag{2.36}$$

In this relation, k is the electron wave vector, $\hbar\omega$ is the phonon energy, and n_w is the average number of phonons of wave vector w; the unprimed and primed quantities refer to the initial and final states respectively. Conservation of energy is ensured by the δ-function in (2.35), and the square of the matrix element is given by

$$|M|^2 = G(w)\begin{cases} n_w \\ n_w + 1 \end{cases} \tag{2.37}$$

for the cases in which the wave number is conserved, i.e.

$$\text{or} \qquad \left.\begin{aligned} k + w &= k' \\ k &= k' + w \end{aligned}\right\}. \tag{2.38}$$

The function $G(w)$ depends on the interaction potential between the electrons and phonons.

We now define the relaxation time $\tau(E)$ for an electron of energy E subject to collisions with the lattice by

$$k_x/\tau(E) = -(\partial k_x/\partial t)_{coll}$$
$$= -\sum_w \{\Delta k_x^a(w) P_w^a + \Delta k_x^e(w) P_w^e\}, \tag{2.39}$$

where x refers to some specified direction, and $\Delta k_x^{a,e}(\mathbf{w})$ is the mean change in k_x in one collision with the phonon of wave vector $\mathbf{w}$. (The mean referred to is over the azimuthal angle of $\mathbf{w}$ around the direction of $\mathbf{k}$; in all cases the superscripts 'a' and 'e' refer to absorption and emission respectively.) The relaxation time $\tau(E)$ is clearly the time constant that measures the exponential decay of the component of momentum in some fixed direction for an electron of energy E. In the absence of an applied field this component of momentum will decay to zero; however, if there is a field F in the x direction, then the rate of change of the component of electron wave vector due to the field is

$$(\partial k_x/\partial t)_F = eF/\hbar. \qquad (2.40)$$

The condition for a quasi-steady state under the combined influence of the lattice and the applied field is then

$$(\partial k_x/\partial t)_{\text{coll}} + (\partial k_x/\partial t)_F = 0. \qquad (2.41)$$

Using eqns (2.39) and (2.40) in eqn (2.41), we find for the average momentum of the electron in the field direction

$$\hbar k_x = eF\tau(E). \qquad (2.42)$$

The current associated with this electron is therefore

$$j(E) = e^2 F\tau(E)/m^*, \qquad (2.43)$$

and the mobility is

$$\mu(E) = e\tau(E)/m^*. \qquad (2.44)$$

The difficulty has therefore been reduced to a calculation of the summation in eqn (2.39) for the particular model of lattice interaction being considered.

To this end we consider in turn polar and non-polar crystals. The reason for the distinction of treatment lies not only in the different form for the interaction function $G(\mathbf{w})$ of eqn (2.37), but also in the different way in which the conditions for the validity of perturbation theory apply. As previously pointed out, an electron in an ionic lattice experiences the greatest interaction with longitudinal optical modes, since these modes produce an electric polarization of the lattice; the phonon energies of these modes will be approximately independent of wave number and of order of or greater than $k_0 T$ for almost all dielectrics and temperatures. On the other hand, the scattering of a conduction electron in non-polar crystals can be considered as due chiefly to interaction with the acoustical lattice modes; for these modes the phonon energy varies approximately linearly with the wave number and is therefore very small for phonons of low wave number. The validity of perturbation theory requires that the average electron energy change per inelastic collision be much smaller than the electron energy. For polar crystals this condition will be met only by high-energy electrons, while for non-polar crystals it should be valid for electrons of thermal energy.

The interaction constant $G(\mathbf{w})$ has been calculated for two special models of polar crystal. In the continuum model, the crystal is represented by a continuous dielectric medium of volume V, with low-frequency and high-frequency dielectric constants ϵ_s and ϵ_∞, and having a single angular frequency ω for longitudinal polarization waves which is independent of the wave number $\mathbf{w}$. For this case,

$$n_{\mathbf{w}} = \{\exp(\hbar\omega/k_0 T) - 1\}^{-1},$$

also independent of $\mathbf{w}$. It is then easily shown (cf. Fröhlich et al. 1950) that

$$G(\mathbf{w}) = 2\pi e^2 \hbar \omega / V w^2 \epsilon^* \qquad (2.45)$$

where

$$\epsilon^{*-1} = \epsilon_\infty^{-1} - \epsilon_s^{-1}, \qquad (2.46)$$

and that this method of calculation should be satisfactory for polarization waves of long wavelength (i.e. $w \ll a^{-1}$, where a is the lattice constant). In the point-charge model, the lattice is represented by a discrete set of point charges alternately $+e$ and $-e$ with ionic masses M_+ and M_-. Fröhlich (1937) shows that

$$G(\mathbf{w}) = 2\pi^2 e^4 \hbar / a^3 w^2 V M \omega, \qquad (2.47)$$

where M is the reduced mass given by $M^{-1} = M_+^{-1} + M_-^{-1}$. This form of interaction function should be satisfactory for $\omega \sim a^{-1}$.

The interactions (2.45) and (2.47) have the same functional dependence on w, but that due to the point-charge model is stronger by a factor of about three for the typical alkali halide. We can therefore replace (2.45) and (2.47) by

$$G(\mathbf{w}) = G_0 w^2, \tag{2.48}$$

where the constant G_0 is given by

$$G_0 = 2\pi e^2 \hbar \omega / V \epsilon^* \quad \text{(continuum model)},$$

$$G_0 = 2\pi^2 e^4 \hbar / a^3 \, V M \, \omega \quad \text{(point-charge model)}.$$

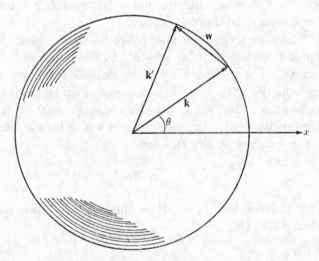

Fig. 2.5. Diagram expressing conservation of wave vector in an electron–phonon collision for the case of absorption. In view of the conservation of energy and momentum, the phonons with which the electron can interact will have wave vectors whose locus will be approximately on a spherical surface, provided the electron energy is much greater than the phonon energy.

Consider now the diagram of Fig. 2.5, which expresses conservation of wave number in the electron–phonon collision for the case of absorption. (For the emission case the phonon wave vector is reversed.) From the diagram,

$$\cos(\mathbf{k}, \mathbf{w}) = -(w^2 + k^2 - k'^2)/2kw,$$

which, together with the conservation of energy

$$\hbar^2 k'^2/2m^* - \hbar^2 k^2/2m^* \pm \hbar\omega = 0,$$

gives

$$\cos(\mathbf{k}, \mathbf{w})^{e,a} = -\frac{w}{2k} \mp \frac{m^*\omega}{\hbar k w}, \qquad (2.49)$$

where the upper and lower signs refer to emission and absorption respectively. If θ is the angle between $\mathbf{k}$ and the x direction, then the required averaging over the azimuth of $\mathbf{w}$ with respect to $\mathbf{k}$ to determine Δk_x gives

$$\frac{\Delta k_x^{e,a}(\mathbf{w})}{k_x} = -\frac{w^2}{2k^2} \mp \frac{m^*\omega}{\hbar k^2}. \qquad (2.50)$$

Using eqns (2.35), (2.37), (2.47), and (2.50) in eqn (2.39) and reducing the sum to an integral (cf. Stratton (1961), Appendix 2), we find

$$\frac{1}{\tau(E, T)} = \frac{Vm^*}{4\pi\hbar^3 k^3}\left\{\int wG_0\left(1 + \frac{2m^*\omega}{\hbar w^2}\right)(1+n_w)\,\mathrm{d}w + \right.$$

$$\left. + \int wG_0\left(1 - \frac{2m^*\omega}{\hbar w^2}\right) n_w\,\mathrm{d}w\right\}, \qquad (2.51)$$

where integration is in each case over the appropriate region of lattice wave numbers. A simplification of eqn (2.51) is possible for the case of high-energy electrons that we are considering. The phonons which lead to large-angle scattering will be those of large wave number (cf. Fig. 2.5), for which $w \sim 1/a$; since $2m\omega a^2/\hbar \ll 1$ we can write (2.51) as

$$\frac{1}{\tau(E, T)} = \frac{Vm^*G_0}{4\pi\hbar^3 k^3}\int w(1+2n_w)\,\mathrm{d}w. \qquad (2.52)$$

The lower limit of w can be put equal to zero in (2.52) since the low wave numbers do not contribute appreciably; the upper limit can be determined from the Debye condition

$$w_0 = (6\pi^2 N)^{\frac{1}{3}}$$

where N is the number of cells per unit volume. For the alkali halides, $N = 1/2a^3$, so that

$$w_0 = (3\pi^2/a^3)^{\frac{1}{3}} \simeq \pi/a. \tag{2.53}$$

Such a large value of w_0 can be used as the upper limit of integration only if the electron interacting with the phonon has sufficient energy to satisfy the requirement of conservation of momentum. Reference to Fig. 2.5 shows that this requires $k_0 > w_0/2$, or in terms of energies $E > E_0$, where

$$E_0 = \hbar^2 w_0^2/8m^* \simeq h^2/32m^*a^2, \tag{2.54}$$

which is an energy of several electronvolts. If the electron energy is lower than the value (2.54), the upper limit of integration will simply by $2k$. Integration of (2.52) then yields

$$\frac{1}{\tau(E,\,T)} = \frac{1}{\tau_0(E)} \left\{ 1 + \frac{2}{\exp(\hbar\omega/k_0 T) - 1} \right\}, \tag{2.55}$$

where the precise form of $\tau_0(E)$ depends on the model chosen to calculate the interaction constant and also on the electron energy. For example, if one uses the point-charge model, and considers an electron whose energy is greater than E_0 (given by eqn (2.54)), one easily obtains

$$\frac{1}{\tau_0(E)} = \frac{\pi^3}{8\sqrt{(2m^*)}} \frac{e^4\hbar}{Ma^5\omega} E^{-\frac{3}{2}}. \tag{2.56}$$

For a lower-energy electron and the continuum model the result is

$$\frac{1}{\tau_0(E)} = \left(\frac{m^*}{2E}\right)^{\frac{1}{2}} \frac{e^2\omega}{\hbar\epsilon^*}. \tag{2.57}$$

The current and the mobility of the electron considered follow from eqns (2.43) and (2.44).

For non-polar crystals, the treatment of the electron–lattice interaction is based on the work of Seitz (1948) in which the interaction energy is introduced in the same way as in the conduction theory of metals, using the deformable-atom hypothesis of Bloch. The scattering of the electron is considered to be due to interaction with the acoustical lattice modes only and an approximate linear relation between frequency and wave number is used,

$$\omega = ws, \qquad (2.58)$$

where s is the velocity of sound. The interaction constant $G(\mathbf{w})$ is found to be given by

$$G(\mathbf{w}) = \frac{2}{9} C^2 w \frac{\hbar}{MNVs}, \qquad (2.59)$$

where C is an energy (of order of 1 eV for many non-polar crystals). If we confine our attention to low-energy electrons (or order of several $k_0 T$) then the conservation of wave number will permit interaction with phonons of low wave number only. Again the collisions will be approximately elastic and Fig. 2.5 will represent the situation with regard to conservation of wave number. For phonons for which eqn (2.58) holds,

$$n_{\mathbf{w}} = \{\exp(\hbar ws/k_0 T) - 1\}^{-1} \simeq k_0 T/\hbar ws, \qquad (2.60)$$

if $\hbar ws \ll k_0 T$, which is true in the case of low wave number. Integration of eqn (2.52) then yields

$$\frac{1}{\tau(E, T)} = \frac{4\sqrt{2}}{9\pi} \frac{C^2 k_0 T m^{*\frac{3}{2}}}{\hbar^4 s^2 MN} E^{\frac{1}{2}}. \qquad (2.61)$$

It should be noted that the use of perturbation theory is well justified for the case of slow electrons in non-polar crystals, but that the value of the constant C is uncertain.

(b) Band conduction due to an electron distribution

So far we have dealt only with the conduction properties of a single electron of specified energy E. We now consider the conduction properties of a distribution of electrons, which, for some unspecified reason, is assumed to be Maxwell–Boltzmann at temperature T (which is the same as the lattice temperature) in the absence of an applied field. Our approach is to calculate the current density j from first principles, and then by analogy from (2.43) and (2.44) to define a mean relaxation time for the distribution, given by

$$\langle \tau \rangle = jm^*/ne^2 F, \qquad (2.62)$$

where n is the electron density in the conduction band, and hence to define a mean mobility

$$\langle \mu \rangle = e\langle \tau \rangle/m^*. \qquad (2.63)$$

Note that $\langle \tau \rangle$ and $\langle \mu \rangle$ are not statistical averages of $\tau(E)$ and $\mu(E)$ but are derived from their defining equations. The definitions lead of course to the current density

$$j = ne\langle \mu \rangle F. \qquad (2.64)$$

In the presence of an external field $\mathbf{F}$, the electron distribution function $f(E, \theta, F)$ will depend on the angle θ between the field direction and the electron momentum $\mathbf{p}$. Let us expand f in a series of Legendre functions,

$$f(E, \theta, F) = f_0(E, F) + f_1(E, F)\cos \theta + \dots$$

To find the relation between f_0 and f_1, we assume that the field causes asymmetry simply by the shift of concentric shells in momentum space by a distance $\Delta p(E) = eF\tau(E)$, where the electrons in the shell have energy $E = p^2/2m^*$. This assumption is reasonable if $\Delta p \ll p$, and this is certainly true for low field strengths. We have then for the asymmetrical part of the distribution function

$$f_1 \cos \theta = - \frac{\partial f_0}{\partial E} \Delta E(E)\cos \theta, \qquad (2.65)$$

where $\Delta E(E)$ is the shift in energy space corresponding to the shift Δp in momentum space. Thus

$$\Delta E(E) = [\{p + eF\tau(E)\}^2 - p^2]/2m^* \simeq (2E/m^*)^{\frac{1}{2}}eF\tau(E), \quad (2.66)$$

so that using eqns (2.65) and (2.66) we have for the coefficient of $\cos \theta$ in the Legendre series expansion

$$f_1(E) = - \left(\frac{2E}{m^*}\right)^{\frac{1}{2}} eF\tau(E) \frac{\partial f_0}{\partial E}. \qquad (2.67)$$

The current density will then be given by

$$j = \frac{1}{3} \int_0^\infty e \frac{p}{m^*} g(E)f_1(E) \, dE, \qquad (2.68)$$

where the factor $\frac{1}{3}$ arises from averaging $\cos^2 \theta$ over all directions in space. Using eqns (2.33), (2.62), and (2.67) in (2.68), one finds

$$\langle\tau\rangle = \frac{\{(2m^*)^{\frac{3}{2}}/(3\pi^2\hbar^3)\} \int\limits_0^\infty \tau(E)E^{\frac{3}{2}}(\partial f_0/\partial E)\, \mathrm{d}E}{\int\limits_0^\infty g(E)f_0(E)\, \mathrm{d}E} \qquad (2.69)$$

which correctly expresses the average value of τ for any form of the symmetrical part of the distribution function. If one assumes that the symmetrical part of the distribution function remains Maxwellian (at least for low field strengths), substitution of $f_0 = \exp(-E/k_0T)$ in eqn (2.69) yields at once

$$\langle\tau\rangle = \frac{4}{3\pi^{\frac{1}{2}}}(k_0T)^{-\frac{5}{2}}\int\limits_0^\infty \tau(E)E^{\frac{3}{2}}\exp(-E/k_0T)\, \mathrm{d}E. \qquad (2.70)$$

We have previously considered the calculation of $\tau(E)$ for electrons in both polar and non-polar crystals. The perturbation calculations are valid only if the electron energy greatly exceeds the phonon energy; for polar crystals, this will require $k_0T \gg \hbar\omega$, which usually requires a temperature much higher than room temperature, since the energy of the optical phonons is approximately constant. For acoustic modes the phonons of low wave number also have low energy, so that for non-polar crystals the approach should be valid, and one finds with the use of eqns (2.61) and (2.70)

$$\langle\tau\rangle = \frac{3\pi^{\frac{1}{2}}}{\sqrt{2}}\frac{\hbar^4 s^2 MN}{C^2(m^*k_0T)^{\frac{3}{2}}}. \qquad (2.71)$$

The mean mobility is then found from eqn (2.63), and the conduction properties of the model are completely determined.

The most frequent applications of this result are to substances that would normally be called semiconductors. However some results on solid argon given by Miller, Howe, and Spear (1968) are interesting in this connection. The mean low-field mobility is shown as a function of temperature in Fig. 2.6;

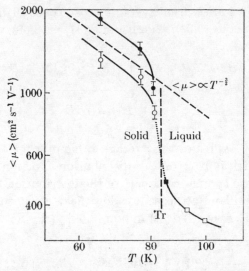

FIG. 2.6. The temperature dependence of the mean low-field mobility in solid argon (after Miller, Howe, and Spear 1968).

the results are consistent with a $T^{-\frac{3}{2}}$ temperature variation of $\langle\mu\rangle$, which is demanded by the calculation on acoustic-mode scattering.

For polar crystals one can properly apply this theory only if $k_0T \gg \hbar\omega$, which is a somewhat unrealistic situation for most polar dielectrics. However, the use of eqns (2.55) and (2.57) in eqn (2.70) yields

$$\langle\tau\rangle = \frac{4\sqrt{2}}{3\pi^{\frac{1}{2}}} \frac{\hbar^2\epsilon^*}{e^2\sqrt{m^*}} \frac{1}{(k_0T)^{\frac{1}{2}}}\,, \qquad (2.72)$$

where we have used the assumption $k_0T \gg \hbar\omega$ to give the approximation

$$\left\{1 + \frac{2}{\exp(\hbar\omega/k_0T) - 1}\right\} \simeq 2k_0T/\hbar\omega. \qquad (2.73)$$

In spite of the fact that perturbation theory is invalid in this connection, it is of interest to find the result that it would yield if applied to the calculation of the relaxation time when $k_0T \lesssim \hbar\omega$. In this case, most electrons will have insufficient energy to emit an optical phonon. Only absorption processes will be allowed unless the electron in question has just absorbed a phonon. Since the absorption probability is proportional to n_w and the emission probability to (n_w+1), such an electron would emit a phonon quickly after absorption. This two-phonon resonance process will therefore occur with a probability approximately equal to that for absorption alone. As far as temperature dependence is concerned, only absorption processes contribute, and the relaxation time will be inversely proportional to n_w. Fröhlich (1954) has shown that simple assumptions lead to an energy-independent mobility for a single electron whose energy is sufficiently low ($E \ll \hbar\omega$); the average mobility for a distribution for which $k_0 T \ll \hbar\omega$ does not differ from this single-electron result. It is given by

$$\langle \tau \rangle = \frac{1}{\sqrt{2}} \frac{\hbar \epsilon^*}{e^2 \omega \sqrt{m^*}} (\hbar\omega)^{\frac{1}{2}} \{\exp(\hbar\omega/k_0 T) - 1\}. \qquad (2.74)$$

It is interesting to note that if the energy-dependent relaxation time given by eqns (2.55) and (2.57) is used in (2.69) one obtains

$$\langle \tau \rangle = \frac{8\sqrt{2}}{3\pi^{\frac{1}{2}}} \frac{\hbar \epsilon^*}{e^2 \omega \sqrt{m^*}} (k_0 T)^{\frac{1}{2}} \{\exp(\hbar\omega/k_0 T) - 1\}. \qquad (2.75)$$

In all cases the mobility is found from the relaxation time with the use of eqn (2.63).

In the section on polaron conduction it will be pointed out that in certain approximations eqn (2.75) needs to be modified only by order-of-magnitude factors to yield a correct result. Discussion of experimental results on polar crystals will be deferred until that section.

(c) Hopping conduction

Difficulties arise with the band model–perturbation theory of electronic conduction if the mobility is very low; there may be various reasons for this failure, but the ultimate experimental manifestation of the situation is a low mobility.

In order to look at some of the reasons for this failure of the usual conductivity theory, we consider the relation between the energy and the wave number in various approximations. For simplicity we restrict ourselves to a one-dimensional lattice with lattice spacing a; if we disregard for the moment any polarization effects, the tight-binding approximation yields for the electron energy as a function of the wave vector,

$$E(k) = 2J(1 - \cos ka), \qquad (2.76)$$

where J is the overlap integral (cf. Ziman 1964). The band width of the band described by eqn (2.76) is given by

$$\Delta W = 4J. \qquad (2.77)$$

For small values of k, eqn (2.76) is equivalent to the quasi-free-electron result (2.32) with the effective mass given by

$$m^* = \hbar^2/2Ja^2. \qquad (2.78)$$

We shall regard the band as broad if $\Delta W > k_0 T$, and narrow in the opposite case. This is illustrated in Fig. 2.7, which shows the energy–wave-number relationship for both broad

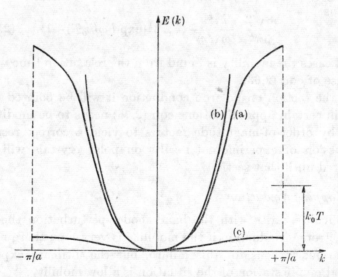

FIG. 2.7. Schematic illustration of broad and narrow energy bands for a one-dimensional lattice. (a) Broad energy band. (b) Quasi-free-electron approximation corresponding to the broad band. (c) Narrow energy band.

and narrow bands, together with the quasi-free electron case corresponding to the broad band.

Consider first the broad-band case, and the restrictions placed on the usual theory by the uncertainty principle. To proceed with order-of-magnitude arguments, we identify the relaxation time τ with the time that a given electron remains in an identifiable plane-wave state (for precise definition of τ see eqn (2.39)); the energy uncertainty is then $\hbar/\tau$. If the distribution of electrons is Maxwellian with temperature T, then the mean electron kinetic energy will be $3k_0T/2$, and the validity of perturbation theory will require the condition

$$3k_0T/2 > \hbar/\tau. \qquad (2.79)$$

With the use of eqn (2.44) the condition (2.79) becomes

$$\mu > 2\hbar e/3m^*k_0T$$

or

$$\mu > 30\frac{m}{m^*}\frac{300}{T}. \qquad (2.80)$$

It should be emphasized that this condition relates to the validity of the perturbation-theory approach provided that the energy band is sufficiently broad compared with k_0T. This condition that the energy bands be sufficiently broad is equivalent to a condition that the effective mass be sufficiently small, as can be seen from eqn (2.78). Thus if one requires $\Delta W > k_0T$ for the validity of the wide-band picture, the use of eqns (2.77) and (2.78) yields the equivalent inequality

$$m^*/m < 2\hbar^2/a^2mk_0T. \qquad (2.81)$$

(At room temperature with $a \sim 3$ Å the condition becomes $m^*/m < 40$.)

Turning to the opposite case of a narrow band in which the concepts of the plane-wave approximation do not apply at all, we see that the uncertainty relation will require the band width to be greater than the uncertainty in the energy of the charge carrier. We can write this

$$\Delta W > \hbar/\tau, \qquad (2.82)$$

where τ is the mean time for the momentum of a charge carrier in any direction to decrease to $1/e$ of its value. The mobility in this case is given by the Einstein relation

$$\mu = eD/k_0T, \tag{2.83}$$

where D is the diffusion coefficient. It is related to the relaxation time by

$$D = \overline{v^2\tau}, \tag{2.84}$$

where the bar indicates an average over the states of the band. The mean value of v^2 may be found from eqn (2.76),

$$\overline{v^2} = \frac{a}{2\pi} \int\limits_{-\pi/a}^{\pi/a} \left\{\frac{1}{\hbar}\left(\frac{\partial E}{\partial k}\right)\right\}^2 \mathrm{d}k$$

$$= (\Delta W)^2 a^2 / 8\hbar^2. \tag{2.85}$$

If the value of τ is taken as constant, the inequality (2.82) reduces to

$$\mu > \frac{ea^2}{8\hbar}\frac{\Delta W}{k_0T} \tag{2.86}$$

with the help of eqns (2.83), (2.84), and (2.85). Slightly different values of the numerical constant appearing in (2.86) are given by Fröhlich and Sewell (1959) and also by Bosman and van Daal (1970); this results from different approaches to an order-of-magnitude estimate. (With $a \sim 3$ Å the value of the constant mobility $ea^2/8\hbar$ is approximately $0\cdot2$ cm^2 v^{-1} s^{-1}.)

To summarize, we have found lower limits for the mobility in a band model for both the broad-band case (equation 2.80)) and the narrow-band case (equation (2.86)). However, there is a difference in the applicability of these conditions, since, as already pointed out, the broad-band criterion refers not only to the validity of the band concept but also to the applicability of perturbation theory to scattering by a single phonon. On the other hand, for the case of narrow bands, the absorption or emission of single phonons becomes impossible in view of energy and wave-vector conservation, so that the relaxation time τ in eqn (2.82) refers to scattering processes other than single-phonon emission or absorption. If there is phonon scattering in the narrow-band case it must be a multiple-phonon process; it is then preferable to drop the idea of a band and to consider the electronic motion as being a thermally activated process in which the electron (described initially by a localized wave function) is displaced to an adjacent localized state. This process has been widely referred to as 'hopping conduction'.

There have been various models used to calculate electron (or hole) mobility. The simplest analysis of experimental results is simply to fit an equation of the type (1.2), and derive a conduction activation energy. Detailed calculations on more or less sophisticated models fit naturally into the following sections on the polaron and impurity conduction; they will therefore be discussed there.

2.3. The polaron

It has been pointed out in the previous section that the interaction between a slow electron and the vibrational modes of a polar lattice is too strong to be treated by perturbation theory. A new approach to the problem was given by Fröhlich *et al.* (1950), who applied the methods of quantum field theory to this problem of solid-state physics. The essential physical idea is that the polarization of the lattice that the electron causes acts back on the electron itself and reduces its energy. As the electron moves through the polar lattice it carries with it this polarization field; the two combined, i.e. the electron and the accompanying polarization field, can be considered as a quasi-particle, which is called a polaron. Since the original paper there has been a vast amount of theoretical work on the polaron, although unambiguous experimental evidence of its existence is less readily available. A very comprehensive account of both theoretical and experimental research has been presented by Appel (1968); Austin and Mott (1969) have given a very readable review of the simple physical basis of the theory of the small polaron, with particular application to TiO_2 and NiO, while Bosman and van Daal (1970) and Adler (1968) have both emphasized applications to transition-metal oxides.

Earlier reviews on the subject of the motion of electrons in polar lattices have been given by Fröhlich (1954) and Allcock (1956). It is not our purpose to present a detailed review of polaron theory, but to extract those portions that may be most useful in helping with the understanding of conduction and breakdown in dielectrics.

The particular model chosen for the calculation of polaron properties depends on whether one is dealing with a large or a small polaron. The polaron size is measured by the extent of the electron wave function and its accompanying lattice distortion; if this is larger than a lattice constant one refers to the polaron as large, otherwise it is called small. For both the large and small polarons the Hamiltonian is written quite generally as

$$H = H_{el} + H_{osc} + H_{int}, \qquad (2.87)$$

exactly in the same form as (2.34), which served as the basis for the perturbation-theory approach to electrical conductivity. The difference is that in polaron theory H_{int} is so large that it must be included in the determination of the zero-order wave functions; it cannot be treated as a small perturbing interaction between the lattice oscillators and the electron. Granted that the problem of finding the zero-order eigenfunctions has been solved, one is then left with the difficulty of providing a theoretical description of the scattering in order to proceed with the calculation of the mobility. The methods used depend not only on whether the polaron is large or small, but also on the strength of the interaction; they vary from perturbation approaches based on a Boltzmann equation formalism to treatments based on the density matrix and the Kubo formalism.

(a) The Fröhlich polaron

The Hamiltonian for the large-polaron problem was derived by Fröhlich (1954), using the language of quantum field theory. The Fröhlich Hamiltonian contains two important characteristic constants, the first of which is of the dimensions of an

inverse length,
$$u = (2m^*\omega/\hbar)^{\frac{1}{2}}. \tag{2.88}$$

In this definition, ω is the angular frequency of the lattice oscillators, and m^* the rigid-band effective mass of the electron, which accounts for the periodicity of the lattice when the ions are fixed at their equilibrium positions. (The electronic polarization of the ions follows the motion of a slow electron adiabatically, and the effect of this part of the polarization is therefore included in the rigid-band effective mass m^*.) The length u^{-1} appears in the theory as a measure of the size of the polaron. If one takes $m^* \sim m$ and $\omega \sim 3 \times 10^{13}$ s^{-1} (which are typical values for an electron in the conduction band of an alkali halide) then $u^{-1} \sim 10$ Å, justifying the use of large-polaron theory in this case.

The second important constant of the theory is the dimensionless quantity
$$\alpha = e^2(m^*/2\hbar^3\omega)^{\frac{1}{2}}/\epsilon^*, \tag{2.89}$$

which is a measure of the strength of the interaction. The effective dielectric constant ϵ^* is defined by eqn (2.73). For an electron in the conduction band of an alkali halide one readily finds $\alpha \sim 5$; for such a value of the coupling constant one speaks of intermediate coupling. If $\alpha < 1$ the situation is referred to as weak coupling, while for $\alpha \gtrsim 10$ the coupling is strong. For a list of polaron coupling constants see Appel (1968). It is clear that the rigid-band effective mass m^* plays a determining role, other things being more or less equal; large values of m^* imply a small polaron and strong coupling, while small values of this quantity lead to large, weakly coupled polarons. It should of course be borne in mind that there are no sharp dividing lines, and each case needs to be treated on its merits.

The first part of polaron theory, viz. the problem of calculating the eigenstates, has been treated by perturbation theory and by variational procedures for weak and intermediate coupling (cf. Fröhlich 1954, Gurari 1953, and Lee, Low, and Pines 1953). The results obtained by these various methods are in substantial agreement and can be summarized as follows.

(1) For low values of the wave number, the energy eigen-values can be written

$$E(k) = E_{\mathrm{p}}(0) + (\hbar^2/2m_{\mathrm{p}}^*)k^2. \qquad (2.90)$$

This expression replaces the quasi-free-electron result (2.32).

(2) The energy of the electron at rest is lowered by the self-energy

$$E_{\mathrm{p}}(0) = -\alpha\hbar\omega. \qquad (2.91)$$

(3) The polaron effective mass is given by

$$m_{\mathrm{p}}^* = m^*(1 + \alpha/6). \qquad (2.92)$$

In attempts to calculate the mobility due to interaction with optical phonons by perturbation methods it is not clear how one should write the interaction that is responsible for scattering the polaron, since at least part of the electron–phonon inter-action has already been included in the theory leading to the formation of the polaron. If one simply ignores this difficulty and regards the scattering of the polaron as being due to interaction with phonons (no corrections being applied), then the theory obtained should be correct in the weak-coupling limit. For the high-temperature case the result is

$$\langle\mu\rangle = (4/3\pi^{\frac{1}{2}})(e/m_{\mathrm{p}}^*\alpha\omega)(\hbar\omega/k_0 T)^{\frac{1}{2}}, \qquad (2.93)$$

which agrees with the mobility derived from eqns (2.63) and (2.72) if the rigid-band effective mass m^* is replaced by the polaron mass m_{p}^* given by (2.92). For the low-temperature case, Howarth and Sondheimer (1953) have found results analogous to (2.75). They find for the mobility

$$\langle\mu\rangle = (8/3\pi^{\frac{1}{2}})(e/m_{\mathrm{p}}^*\alpha\omega)(k_0 T/\hbar\omega)^{\frac{1}{2}} \times$$
$$\times G(T)\{\exp(\hbar\omega/k_0 T) - 1\}, \qquad (2.94)$$

where $G(T)$ is a slowly varying function of temperature of order of magnitude unity. The inclusion of this factor and the replacement of m^* by m_{p}^* are the only differences between (2.94) and the result obtained by use of (2.75) in (2.63). Other approaches to the low-temperature scattering of polarons by optical lattice modes are given by Appel (1968); in all cases the temperature dependence of the mobility is dominated by the exponential term, which was first found by Fröhlich (1937).

Experimental verification of large-polaron theory in the weak- and intermediate-coupling case has concentrated on the temperature dependence of the theoretical result as applied to alkali halides. The method used in the experimental work has been that of Redfield (1954), who measured the Hall mobility of photoelectrons; the photo-generation of electrons avoids problems with electrode contacts, and the Hall mobility measurements are true mobilities, as distinct from drift mobilities, which could be trap-controlled in the alkali halides.

Brown and Inchauspé (1961) and Ahrenkiel and Brown (1964) have measured the Hall mobilities of various alkali halides over a temperature range of 7–100 K; their results are shown in Figs. 2.8–2.10. The general features of these curves are illustrated by the construction on Fig. 2.9. Below about 20 K the results are adequately explained by a constant mobility due to scattering by defect centres, and above 40 K the data are well fitted by an equation of the form (2.94) in which the scattering is due to polar optical modes. However, it was necessary to introduce the assumption that appreciable scattering by acoustic modes existed in the temperature range between 20 K and 40 K.

Seager and Emin (1970) have continued the Hall mobility measurements to higher temperatures. Figures 2.11–2.14 show these results taken on various alkali halides over the temperature range 160 K to 400 K. In each case the dashed line corresponds to the Howarth and Sondheimer (1953) theory, with the temperature-independent part of eqn (2.94) adjusted to fit the data at the lowest temperatures for which measurements were taken. Using data for the alkali halides (Appendix 1) one finds that this fit requires values of the polaron mass of order of or slightly greater than the free-electron mass. The experimental results are in good agreement with the theory, especially those for NaCl and KI. The theoretical mobility (2.94) has been derived on the assumption of a low temperature, i.e. $T < \hbar\omega/k_0$; this is certainly not true for the case of caesium

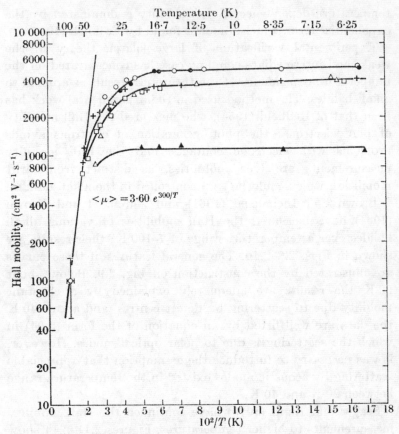

FIG. 2.8. Low-field Hall mobility of photoelectrons in KCl as a function of temperature for different F-centre concentrations as follows: ● 1.5×10^{15} cm^{-3}, ○ 2.8×10^{15} cm^{-3}, △ 5.3×10^{15} cm^{-3}, $+$ 2.4×10^{16} cm^{-3}, □ 5.6×10^{16} cm^{-3}, ▲ 3.4×10^{17} cm^{-3}. The straight line shows the exponential variation of mobility corresponding to optical phonon scattering. (After Brown and Inchauspé 1961.)

bromide. This case has been discussed by Seager and Emin (1970), and we shall return to it below in connection with small polarons.

The electron drift mobility in KCl has been measured by Hirth and Tödheide-Haupt (1969), who used highly zone-refined material in order to minimize the effect of electron trapping. Their results are shown in Fig. 2.15; above 300 K these values

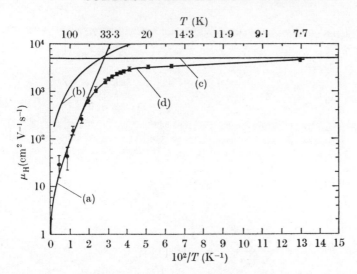

FIG. 2.9. Low-field Hall mobility of photoelectrons in KI as a function of temperature. (a) $\langle\mu_{op}\rangle = 10\{\exp(222/T)-1\}$ (cf. eqn (2.94)). (b) $\langle\mu_{ac}\rangle = 1\cdot15\times10^6 T^{-\frac{3}{2}}$ (cf. eqn (2.71)). (c) $\langle\mu_e\rangle = 5000$. (d) Total mobility $\langle\mu_H\rangle$ given by $\langle\mu_H\rangle^{-1} = \langle\mu_{op}\rangle^{-1}+\langle\mu_{ac}\rangle^{-1}+\langle\mu_e\rangle^{-1}$. After Ahrenkiel and Brown (1964).

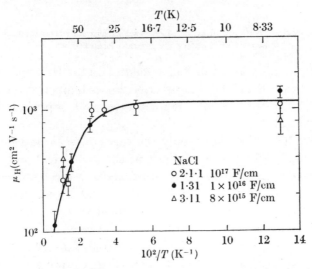

FIG. 2.10. Low-field Hall mobility of photoelectrons in NaCl as a function of temperature for various F-centre concentrations. (After Ahrenkiel and Brown 1964). The point marked □ is due to Redfield (1954).

3

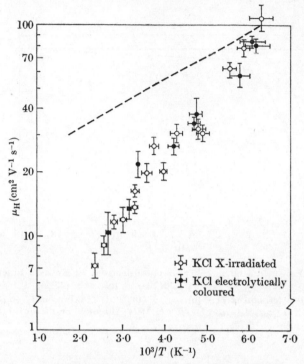

FIG. 2.11. Low-field Hall mobility of photoelectrons in KCl; the dashed line
is equation (2.94) (after Seager and Emin 1970).

of the drift mobility are in fair agreement with the Hall mobility
results shown in Fig. 2.11. However, at lower temperatures the
drift mobility increases with temperature, and is presumably
trap-controlled.

So far we have considered only the cases of weak and intermediate coupling for the large polaron; the theoretical problem
of the strongly coupled large polaron has also received much
attention. However, it has already been pointed out that the
factors that make polarons strongly coupled also tend to make
them small. If one sets $\alpha = 20$ and uses typical alkali halide
data $\epsilon^* \sim 5$ and $\omega \sim 3 \times 10^{13}\,\text{s}^{-1}$ then from eqns (2.88) and
(2.89) $m^* \sim 12m$ and $u^{-1} \sim 3\,\text{Å}$. Strong coupling in the
alkali halides is therefore more reasonably treated as a small-
polaron problem.

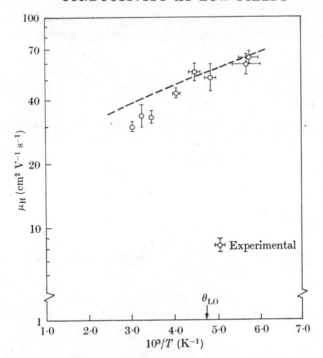

FIG. 2.12. Low-field Hall mobility of photoelectrons in KI; the dashed line is eqn (2.94) (after Seager and Emin 1970).

For the sake of completeness we simply state the appropriate results for the large strongly coupled polaron as found by Feynman (1955) using a path-integral method. He found for the ground-state energy at zero temperature

$$E_p(0) = -(0.106\alpha^2 + 2.83)\hbar\omega, \qquad (2.95)$$

and for the polaron effective mass

$$m_p^* = (16\alpha^4/81\pi^4)m^* \simeq (\alpha^4/500)m^*. \qquad (2.96)$$

The mobility has been calculated by Feynman, Hellwarth, Iddings, and Platzman (1962); the expression is not a simple one, but for low temperatures and strong coupling it is found that

$$\langle\mu\rangle \propto (e/m_p^*\alpha\omega)(k_0T/\hbar\omega)\alpha^{-2}\exp\{(\hbar\omega/k_0T) + \alpha^2\}. \qquad (2.97)$$

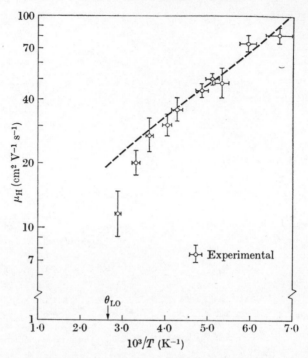

Fɪɢ. 2.13. Low-field Hall mobility of photoelectrons in NaCl; the dashed line is eqn (2.94) (after Seager and Emin 1970).

The temperature dependence of this result is only very slightly different from that of (2.94), but the dependence on coupling constant is very strong in (2.97).

(b) *The small polaron—Holstein's model*

In the large Fröhlich polaron the polarization of the lattice is treated by a continuum approximation, and the electron is introduced into the basic Hamiltonian as a free particle of mass m^*. However, if the characteristic length u^{-1} (cf. eqn (2.88)) is of the order of the lattice spacing, it is no longer reasonable to use the continuum approximation in calculating the polarization of the lattice; moreover in this case it is better to use the Heitler–London approach to electrical conductivity, and select localized orbitals for the zero-order electron wave functions.

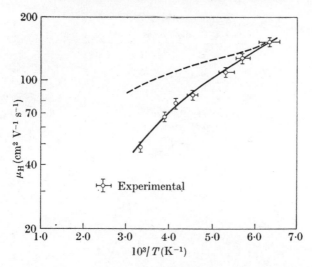

Fig. 2.14. Low-field Hall mobility of photoelectrons in CsBr. Both the dashed and the solid lines refer to eqn (2.94); in the former case the value of m_p is taken from (2.92) while in the latter case (2.102) is used (after Seager and Emin 1970).

The coupling constant α, which has played a leading role in the theory of the Fröhlich polaron, will need to be replaced by a quantity that is in accord with the basic assumptions. In particular it will not contain the dielectric constants ϵ_0 and ϵ_∞, which relate to continuum polarization theories; instead there will be sums over the vibrational modes of the distorted lattice. The definition of such a coupling constant is cumbersome and its evaluation requires some form of computation. We therefore adopt the rather roundabout procedure of using a definition introduced by Holstein (1959), evaluated in continuum approximation by Yamashita and Kurosawa (1958) and computed approximately by Bosman and van Daal (1970). This coupling constant is given by

$$\gamma \simeq (1{\cdot}25/\pi)(e^2/\hbar\omega a\epsilon^*),$$

and leads to a maximum polaron binding energy

$$E_p(0) = -\gamma\hbar\omega. \tag{2.98}$$

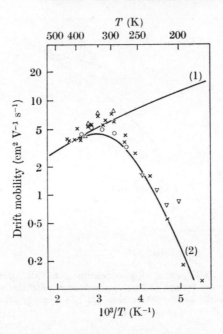

FIG. 2.15. Electron drift mobility in KCl as a function of temperature for four different zone-refined samples. Curve (1) is a microscopic mobility whose temperature dependence is the same as (2.94), and calculated on the assumption that curve (2) represents a trap-controlled drift mobility. (After Hirth and Tödheide-Haupt 1969.)

In this case of narrow bands and strong coupling constants we can consider the electron to be moving so slowly that it becomes trapped in its own polarization field (this idea of self-trapping was introduced by Landau (1933)). The question of polaron mobility then devolves on the analysis of the motion of this self-trapped electron. Two basic processes are possible. The electron may move to an adjacent site by quantum-mechanical tunnelling. Should this be the dominant transport process, conduction will be of the band type. On the other hand, phonon-induced transitions to adjacent sites may occur more frequently than tunnelling. This process (already mentioned in §2.2(c)) is called hopping conduction; clearly it will be favoured by high temperatures.

The criterion for distinguishing between these two conduction processes has been worked out by Holstein (1959) on the basis of a special model of a linear chain of polar molecules having their orientations and mass centres fixed. He finds that the transition from band to hopping behaviour should take place at a temperature T_t given by

$$T_t \simeq \hbar\omega/2k_0. \qquad (2.99)$$

Let us consider first small-polaron band conduction. Holstein (1959) finds that the width of the small-polaron conduction band is given by

$$\Delta W_p = 4J \exp\{-S(T)\}, \qquad (2.100)$$

where

$$S(T) = \gamma \coth(\hbar\omega/2k_0 T). \qquad (2.101)$$

Comparison with eqns (2.76), (2.77), and (2.78) shows that we can define the effective small-polaron mass by

$$m_p^* = m^* \exp\{S(T)\}, \qquad (2.102)$$

where m^* is the rigid-lattice effective mass given by (2.78). Clearly the band width is a strongly decreasing function of temperature and the effective mass a strongly increasing function of temperature. For the average band mobility Holstein (1959) finds

$$\langle\mu\rangle = (ea^2\omega/\pi^{\frac{1}{2}}k_0 T)\{\gamma \operatorname{cosech}(\hbar\omega/4k_0 T)\}^{\frac{1}{2}} \times$$
$$\times \exp\{-2\gamma \operatorname{cosech}(\hbar\omega/2k_0 T)\}. \quad (2.103)$$

There seems to be no experimental evidence for the actual existence of small-polaron band conduction. However, in an effort to make a better fit with experimental results, Seager and Emin (1970) have tentatively used the result (2.94) for the mobility with (2.102) used as the polaron mass rather than (2.92). It should be noted that their experimental results are for temperatures well above T_t of (2.99); this drawback is aggravated by the fact that T_t is believed to be strictly an upper limit (see Bosman and van Daal (1970)).

We turn now to the case of hopping conduction, for which the calculation of the mobility was first carried out by Yamashita and Kurosawa (1958); their result is

$$\langle \mu \rangle = \frac{ea^2}{k_0 T} \frac{J^2}{\hbar^2 \omega} \left\{ \frac{\pi \sinh(\hbar\omega/4k_0 T)}{\gamma} \right\}^{\frac{1}{2}} \times$$

$$\times \exp\{-2\gamma \tanh(\hbar\omega/4k_0 T)\}, \quad (2.104)$$

which was also found by Holstein (1959) for his linear-chain model. For high temperatures, eqn (2.104) can be approximated by

$$\langle \mu \rangle = \mu_0 \exp(-W_H/k_0 T), \quad (2.105)$$

where μ_0 varies slowly with temperature and is given by

$$\mu_0 = \frac{ea^2}{k_0 T} \frac{1}{\hbar} \left(\frac{\pi}{4\gamma\hbar\omega} \frac{1}{k_0 T} \right)^{\frac{1}{2}} J^2, \quad (2.106)$$

and the activation energy for hopping is given by

$$W_H = \gamma\hbar\omega/2. \quad (2.107)$$

The results quoted above for small-polaron hopping conduction all refer to the so-called non-adiabatic case; although the motion of the polaron in its self-created polarization well is adiabatic, the tunnelling probability during an excited state of the whole system is small. Moreover, eqn (2.104) was derived under mathematical approximations that require both strong coupling and high temperatures; computations on the effect of relaxation of these restrictions have been given by de Wit (1968).

There is unambiguous evidence for the existence of the small polaron as a hole in the valence band of certain alkali halides. Castner and Känzig (1957) discovered such localized holes in LiF, NaCl, KCl, and KBr using electron spin resonance techniques. The experimental results were consistent with localization of the hole on two halide ions, thus forming a tightly bound, singly ionized halogen molecule oriented in the (110) direction. The distance between the halogen nuclei was of order one-half the interionic distance of the perfect lattice; such a system is called a V_K centre. Keller, Murray, Abraham, and Weeks (1967) have investigated the thermal motion of V_K centres in KCl, and they find that the centres reorient through

60° with an activation energy of 0·54 eV. This agrees well with the activation energy for long-range diffusion measured by Neubert and Reffner (1962); it is therefore presumably also the activation energy for hopping conduction. Nettel (1963) has attempted to calculate the ground-state energy of the self-trapped hole in KCl, but the theoretical basis for the calculation is only moderately well supported by the results of the electron spin resonance experiments. Appel (1968) quotes a figure of 0·32 eV for hopping activation energy in LiF and gives an estimate for $\mu_0 \sim 3500 \ \mathrm{cm^2 \ v^{-1} \ s^{-1}}$ at 110 K; this leads to a mean mobility $\langle \mu \rangle \sim 1\cdot7 \times 10^{-14} \ \mathrm{cm^2 \ V^{-1} \ s^{-1}}$ at this temperature.

There are at least two other substances in which electrical conduction has been interpreted in terms of the polaron model, although little or no work exists on their high-field properties; these are sulphur and nickel oxide. Gibbons and Spear (1966) have measured the temperature dependence of electron drift mobility in orthorhombic sulphur crystals, and their results are shown in Fig. 2.16 for the (111) crystallographic direction. The results were the same along other principal directions in the crystal. The mobility is very low over the entire temperature range, and the activation energy derived from the experimental results is 0·167 eV. Gibbons and Spear (1966) rule out trap-controlled transport on the grounds that the results apparently do not depend on crystal purity and manner of activation of the carriers; they therefore conclude that the electron transport is determined by centres whose density is a property of the sulphur crystals. The results were fitted to eqn (2.104) for a selection of values of γ which were large enough to guarantee the linear relation shown in Fig. 2.16 within experimental error. The value for the polaron binding energy was found to be $E_{\mathrm{p}}(0) = 0\cdot48$ eV. Equation (2.98) can then be used to determine $\hbar\omega$ if γ is known. There are several prominent modes in the molecular vibration spectrum of sulphur which fit these results for values of γ between 17 and 27. However, the model on which small-polaron theory is based assigns a single vibrational frequency to the lattice; if the lattice has several

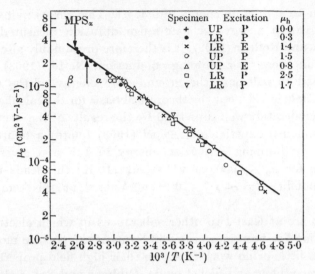

FIG. 2.16. Temperature dependence of the electron drift mobility in the (111) direction in orthorhombic sulphur crystals. LR, crystal grown from laboratory reagents; UP, crystal grown from ultra-pure reagents; E, electron-beam excitation; P, photon excitation; μ_h is the room-temperature hole drift mobility in $cm^2 V^{-1} s^{-1}$. (After Gibbons and Spear 1966.)

vibrational frequencies, they will presumably all contribute to the formation of the polaron. Gibbons and Spear (1966) point out that, in view of this, it is not possible to be definite about assignments of various parameters in the theory.

Electrical conduction in NiO is reviewed in great detail by Bosman and van Daal (1970). Early investigations on p-type NiO by Heikes and Johnston (1957) led to the belief that hole drift mobility was exceedingly low ($\sim 10^{-3}$–10^{-9} $cm^2 V^{-1} s^{-1}$) and exhibited a thermally activated behaviour; this has been questioned by Bosman and van Daal (1970), who cite reasons connected with dope levels and grain boundary layers for believing that neither is the mobility so low (~ 1 $cm^2 V^{-1} s^{-1}$) nor is the transfer process thermally activated. They believe that free charge carriers are well described by the large-polaron model, whereas charge carriers bound to centres have to be considered as small polarons. As mentioned above the break-down and high-field properties of NiO have not been intensively investigated. However, the difficulties encountered in interpreting conduction in NiO are a reminder that current interpretations of conduction properties of many other dielectrics may also undergo very substantial revision; this comment is particularly apposite to high-field conduction processes, for which there is a general scarcity of reliable data.

2.4. Impurity conduction

The phenomenon of impurity conduction in semiconductors is well known and has been reviewed by Mott and Twose (1961), Mott (1967), and Klinger (1968). A qualitative review of the adjacent field of conduction in amorphous semiconductors has been given by Cohen (1971). The basic idea is that electrons occupying isolated donor levels (or holes occupying acceptor levels) may contribute to the electrical conductivity in a manner other than the usual process of thermal excitation followed by band conduction. The electron occupying the isolated donor is described by a localized wave function, which will overlap by a small but finite amount with neighbouring donors. Charge transport can occur by tunnelling to an adjacent vacant donor level and this has been called impurity band conduction. Mott and Twose (1961) object to the use of the term 'band' in this connection, on the grounds that the donor impurities are randomly rather than regularly distributed; this renders many well-known band properties invalid, e.g. the negative effective mass of electrons occupying states near a band maximum. However, the description of this process as impurity band conduction has become general. Alternatively, charge transport may proceed by multiphonon excitation of the localized electron over the potential barrier separating it from an adjacent vacant site; this process can be called impurity hopping conduction.

The characteristic features of impurity conduction, particularly those that distinguish it from polaron conduction, are as follows.

(1) The centres at which the electron wave function can be localized are randomly distributed, even though their average

density may be uniform. If these centres were distributed in a regular periodic array, the conduction would be similar to small-polaron conduction.

(2) There must be an adjacent vacant centre to which an electron can move, and this can arise in two ways. If all of the localized levels for electrons and holes are due to charged centres (donors and acceptors), then impurity conduction can take place only if there is compensation, i.e. the presence of some minority centres. Figure 2.17 illustrates this situation,

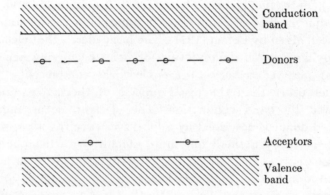

Fig. 2.17. Diagram illustrating partial compensation of n-type material. The horizontal lines represent localized centres, and the circles electrons in them.

in which n-type material is partially compensated, leading to the required vacant donor levels. However, this discussion for the necessity of compensation is based on the notion that trapping centres for electrons and holes are necessarily charged, so that an n-type material containing no minority centres can have no vacant trapping centres for electrons. This may be true for crystalline semiconductors, but for many dielectrics trapping may occur at centres that are uncharged; this should be most marked for amorphous dielectrics.

The whole of this discussion concerning the need for an adjacent vacant centre supposes that the overlap between centres is very small; if it is large a quasi-metallic conductivity ensues, which is not of interest for our purposes.

(3) The energies of the localized charge carriers may not be the same in two adjacent positions, and this may arise for two different reasons. In the first place the impurity centres may be different, and the energies of trapped charge carriers correspondingly different. However, even if the impurity centres are identical, both they and the centres for minority carriers are randomly distributed through the lattice. Owing to this random distribution of positively and negatively charged centres, there is a fluctuation of energy between one centre and the next.

Anderson (1958) used a special model to investigate the effect of random variations in potential on the band structure that would otherwise be associated with an array of centres. He found that when the randomness is sufficiently great, all states are localized; the results as given by Anderson (1958) are in a complicated form, but Mott (1967) estimates that states will be localized if the spread of the random potentials exceeds about six times the band width. On the basis of this theoretical work, Mott (1967) conjectured a model for the band structure of a disordered insulator; the details of the model are as follows. Within a given range of energies, states will either be localized or not; certain critical values of the energy will separate localized from non-localized states. This is shown schematically in Fig. 2.18; states are non-localized for energies higher than E_c in the conduction levels and lower than E_v in

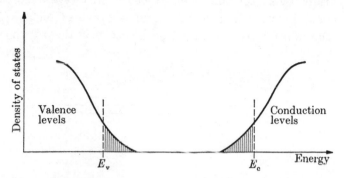

FIG. 2.18. A schematic diagram of the density of states for an insulator with disordered structure; the shaded portions correspond to localized states (after Mott 1967).

the valence levels, while below E_c and above E_v are tails of localized states. For non-localized states the electron mean free path tends towards its de Broglie wavelength as the energy approaches either of the critical values in the valence or conduction levels. For localized states, the wave function falls off exponentially with distance, and the exponent approaches zero as the energy approaches the critical values.

Conduction in this model can occur by either one of two mechanisms or by a combination of both; we shall discuss these mechanisms separately.

(a) Trap-controlled band conduction

Excitation of electrons into the non-localized levels of the conduction states or from the non-localized levels of the valence states results in electron or hole current, as described by the usual band theory of conduction. This process is associated with an activation energy corresponding to the energy required for excitation of the charge carriers into non-localized levels.

In his treatment of intrinsic breakdown in amorphous dielectrics, Fröhlich (1947a) proposed a model that is essentially equivalent to the one under discussion; the Fröhlich model can be used to calculate the temperature dependence of trap-controlled band conduction. The energy-level diagram for the model is depicted in Fig. 2.19. The localized levels below the conduction levels are assumed to be of two types: shallow traps, or S levels, which extend over a range of energy ΔV below the non-localized conduction levels; and deep traps, or

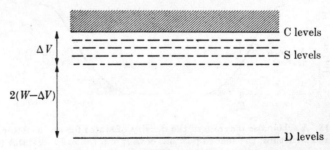

Fig. 2.19. Schematic energy-level diagram for an amorphous dielectric.

D levels, which lie below the C levels by an amount $2\,W \gg \Delta V$. This rather sharp distinction between different types of localized level is not envisaged in Fig. 2.18, but the models are nevertheless qualitatively similar. The electrons in C, S, and D levels are assumed to be sufficiently numerous to establish a thermal equilibrium amongst themselves at some temperature T_{e}, which, in the presence of a strong applied field, may be different from the lattice temperature T.

In addition, the validity of the inequalities

$$k_0 T_{\mathrm{e}} \ll \Delta V \ll W \qquad\qquad (2.108)$$

is required in all cases considered. Given sufficient available traps, (2.108) implies

$$n_{\mathrm{C}} \ll n_{\mathrm{S}} \ll n_{\mathrm{D}}, \qquad\qquad (2.109)$$

where n_{C}, n_{S}, and n_{D} are the numbers of electrons in non-localized conduction levels, shallow traps, and deep traps respectively. In view of the basic assumption of the model that thermal equilibrium is quickly established between C, S, and D electrons, n_{C} and n_{D} will be related by

$$n_{\mathrm{C}} = n_{\mathrm{D}} \gamma \exp(-W/k_0 T_{\mathrm{e}}), \qquad\qquad (2.110)$$

where γ is the ratio of the density of energy levels in a range $k_0 T_{\mathrm{e}}$ near the bottom of the C levels, to the corresponding quantity near the bottom of the D levels.

Using equation (1.1), we have for the conductivity

$$\sigma = n_{\mathrm{C}} e \mu = n_{\mathrm{D}} e \mu \gamma \exp(-W/k_0 T_{\mathrm{e}}), \qquad\qquad (2.111)$$

where μ is the mobility of an electron in a non-localized C state. This mobility is usually temperature-dependent (cf. eqns (2.71) and (2.75)), but, provided that $W \gg k_0 T_{\mathrm{e}}$, the main part of the temperature dependence will be in the exponential term of (2.111). This type of conduction is therefore associated with an activation energy W; for low field strengths the temperature T_{e} of the C, S, and D electrons will closely approximate to the lattice temperature T.

Equation (2.111) is sometimes interpreted in terms of an

effective mobility
$$\mu_{\text{eff}} = \mu\gamma \exp(-W/k_0 T) \qquad (2.112)$$

with the conductivity given by

$$\sigma = n_{\text{S}} e \mu_{\text{eff}}. \qquad (2.113)$$

Since $n_{\text{C}} \ll n_{\text{S}}$, this is approximately equivalent to regarding all of the non-localized and shallow-trapped localized electrons as charge carriers, and the mobility as associated with an activation energy. Whether one adopts the point of view expressed by (2.111) or that of (2.112) and (2.113) is merely a matter of convenience.

The conductivity of various insulating solids has been interpreted in this way by Spear and co-workers (see references below). Drift-mobility measurements were made on carriers generated by a pulse of high-energy electrons; the results of Adams and Spear (1964) for the hole mobility in orthorhombic sulphur crystals are shown in Fig. 2.20. The high-temperature mobility is fitted empirically to a $T^{-\frac{3}{2}}$ law, as described by eqn (2.71) for scattering by acoustic modes of a non-polar crystal; for lower temperatures, eqn (2.112) was fitted. The results were consistent with a trapping level for holes 0·19 eV above the valence band, and a density of trapping levels from 4×10^{14} cm^{-3} to 10^{17} cm^{-3}, as shown on Fig. 2.20. The trapping centres were identified as being due to dissolved CS_2 molecules.

Essentially similar results have been found by Spear (1961) for the electron mobility in both crystalline and vitreous selenium. The trapping centre was 0·25 eV below the conduction levels in both cases; the density of traps was of order 10^{14} cm^{-3} for the crystalline samples and 10^{18} cm^{-3} for the vitreous form.

The electron mobility of high-resistivity cadmium sulphide has been subject to a similar analysis by Spear and Mort (1963), who find the low-temperature conductivity trap-controlled with a trap depth in the range 0·02–0·04 eV, depending on trap concentration. This situation requires a more detailed analysis than the one presented here, since the inequality (2.108) will not be true except at very low temperatures.

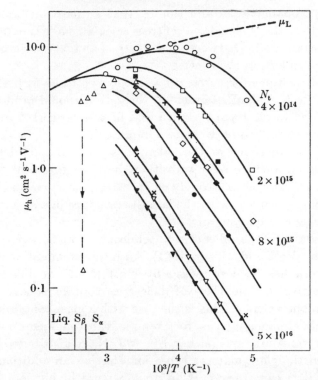

FIG. 2.20. Temperature dependence of the hole drift mobility in orthorhombic sulphur crystals for various trap concentrations (after Adams and Spear 1964).

(b) *Impurity conduction by hopping*

Hopping from one localized level to an adjacent one results in a transfer of charge. This process is also associated with an activation energy, since the energy of a given localized state is in general different from that of an adjacent state. There are two basic problems to consider in evaluation of the conductivity due to such a process; the first is to calculate the hopping probability between centres, and the second is to make an appropriate average, bearing in mind the random distribution of centres.

This problem has been solved by Miller and Abrahams (1960) for a special model of an n-type semiconductor. The steps in their calculation are as follows.

(1) The probability per unit time P_{ij} that an electron jumps from the ith to the jth localized centre is calculated as a function of the distance R_{ij} between the centres and the electron energies E_i and E_j in the centres.

(2) The equations for current flow are shown to be formally equivalent to Kirchhoff's Laws for a three-dimensional random resistance network. Each resistive link in the network corresponds to a transition between two impurity centres.

(3) The conductivity of the network is then computed by performing an average over transitions between the centres i and j; clearly it depends strongly on the path of least resistance for each charge carrier, with the largest resistance in a series of links playing a dominant role.

This work and its application to semiconductors is reviewed in detail by Mott and Twose (1961). An intuitive modification of this work has been proposed by Mott (1967) as possibly applicable to a certain class of dielectrics containing ions of more than one valency. This is the case with a glass containing transition-metal ions such as, for example, vanadium pentoxide glass. The vanadium ions (either V^{4+} or V^{5+}) are distributed at random in the glass matrix; those ions having an additional electron (V^{4+}) can transfer it to an ion more highly ionized (V^{5+}). The Mott formula can be justified as follows.

Consider a simple one-dimensional model containing isolated centres distant R apart, between which an electron can make transitions by hopping motion. If the probability per unit time that the electron makes such a transition in the absence of a field is denoted by P, the presence of an applied field F will modify this probability so that it will be $P \exp(eFR/k_0T)$ in one direction and $P \exp(-eFR/k_0T)$ in the other. Expanding the exponentials to first order, we have for the particle current between two centres $2PeFR/k_0T$. The extension to three dimensions introduces a factor $\frac{1}{3}$ to account for random orientation of the direction of hopping with respect to the applied field, and a factor $1/R^2$ to account for the number of hopping paths across unit area. Let the relative concentration of centres with an excess electron be denoted by c, and the concentration of centres that can accept an electron by $(1-c)$. Ignoring factors of order of magnitude unity, we get for the conductivity

$$\sigma = c(1-c)\frac{e^2}{Rk_0T}\,P. \qquad (2.114)$$

The transition probability between localized centres is written

$$P = vp(R)\exp[-\{\Delta E + \tfrac{1}{2}E_p(0)\}/k_0T], \qquad (2.115)$$

where v is a phonon frequency, $E_p(0)$ is the small-polaron binding energy of eqn (2.98), and ΔE is the average difference in energy between the two types of level. The factor $p(R)$ takes account of tunnelling, and it must be introduced if the distance between the states is large. If the two impurity levels have a mean energy W below the conduction band, then $p(R)$ is given by

$$p(R) = \exp(-2\alpha R) \qquad (2.116)$$

where

$$\alpha = (2mW)^{\frac{1}{2}}/\hbar.$$

Combining eqns (2.114), (2.115), and (2.116), we arrive at the Mott formula

$$\sigma = vc(1-c)(e^2/Rk_0T)\exp(-2\alpha R) \times$$
$$\times \exp[-\{\Delta E + \tfrac{1}{2}E_p(0)\}/k_0T]. \qquad (2.117)$$

Various experimental results are cited by Austin and Mott (1969) in support of formula (2.117) for the conductivity of vanadate glasses. The main points are summarized below.

(1) From measurements of the Seebeck coefficient made by Kennedy and Mackenzie (1967), it is inferred that $\Delta E \ll k_0T$, so that only the polaron energy remains in the conduction activation energy.

(2) Since the static dielectric constant of these glasses is very high (in the range 15 to 50), the small-polaron energy (eqn (2.98)) is inversely proportional to the high-frequency dielectric constant. Experimental results from the work of Nester and Kingery (1963) are shown in Table 2.3; the product of the activation energy for conductivity and the high-frequency dielectric constant is approximately a constant, which is consistent with the idea that the activation energy is a polaron energy.

TABLE 2.3

Data on the electrical conductivity of vanadate glasses of various compositions (after Nester and Kingery (1963)).

Activation energy of conductivity ϕ (eV)	High-frequency dielectric constant ϵ_∞	$\phi\epsilon_\infty$ (eV)
0·295	4·05	1·19
0·333	3·72	1·23
0·392	3·35	1·31
0·418	3·22	1·35
0·443	3·15	1·39

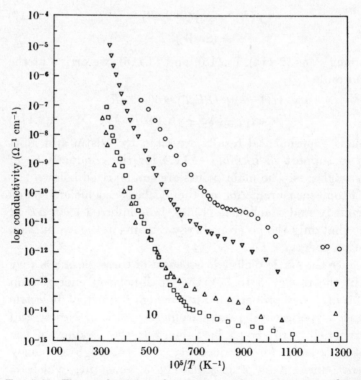

FIG. 2.21. The conductivity of various vanadate glasses as a function of temperature. The mole ratios of $V_2O_5:P_2O_5$ are as follows. △ 1:3, □ 1:1, ▽ 6:1, ○ 7:1. (After Schmid 1968.)

(3) The factor $\exp(-2\alpha R)$ of eqn (2.117) should result in a linear relation between the logarithm of the conductivity and the mean distance between centres. Austin and Mott quote unpublished work which verifies this relation for vanadate glasses.

The conductivity of vanadate glasses has also been studied by Schmid (1968), whose results are shown in Fig. 2.21 for glasses of various compositions. The apparent activation energy of the conductivity is of order 0·3–0·5 eV at high temperatures in agreement with the results of Nester and Kingery (1963) quoted in Table 2.3, but it falls to a much lower value for temperatures below 150 K. This more or less abrupt change is tentatively ascribed to a transition from hopping to band conduction predicted by the Holstein model of the small polaron.

LeComber and Spear (1970) have investigated electronic transport in high-resistivity amorphous silicon films, and their measurements of drift mobility are shown in Fig. 2.22. For

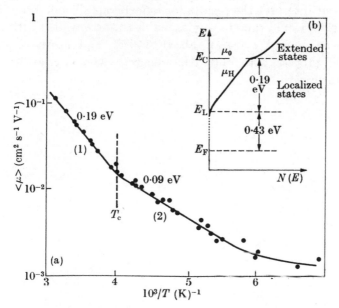

FIG. 2.22. Temperature dependence of the electron drift mobility in an amorphous silicon film, 1·3 μm thick (after LeComber and Spear 1970.)

temperatures above 240 K they interpret their results in terms
of a trap-controlled mobility (eqn (2.112)) with $\Delta V = 0.19$ eV.
Below 240 K they suggest a thermally activated hopping
mechanism, with a mobility activation energy of 0·09 eV. For
the same two regions, the electrical-conductivity activation
energies are 0·62 eV and 0·51 eV respectively, since the number
of carriers is also an exponential function of the temperature.
The importance of measuring simultaneously the activation
energies of both the mobility and the conductivity has been
stressed by Mort (1971), who points out that the activation
energy of a trap-controlled mobility does not manifest itself
in the conductivity. This is true if the carriers in deep traps are
in thermal equilibrium with the carriers in shallow traps and
conduction levels, as is presumably the case in amorphous
silicon. However it will not be true otherwise, as for example
in the Fröhlich model described in § 2.4(a).

It is possible that the results of Schmid for vanadate glass
(Fig. 2.21) can be interpreted in the same way as LeComber
and Spear interpret the results for amorphous silicon. Mobility
in the glass would then be envisaged as trap-controlled with an
activation energy $\sim$0·4 eV above 150 K, and as due to a
hopping process with activation energy $\sim$0·05 eV at lower
temperatures.

HIGH FIELD CONDUCTION:
ELECTRODE EFFECTS

3.1. Fowler–Nordheim emission

FIELD emission, or the quantum-mechanical tunnelling of electrons from a metal surface into a vacuum under the influence of a strong electric field, was first explained by Fowler and Nordheim (1928), whose theory has since been refined in many ways by various authors; reviews of this work are given by Good and Müller (1956) and by Lamb (1967). We propose to set out the main features of field-emission theory and to discuss the modifications appropriate when the emission is into a dielectric rather than a vacuum.

The simplest model for field emission into a dielectric uses a triangular potential barrier (see Fig. 3.1(a)) and assumes that the electron distribution in the metal does not differ significantly from that at the absolute zero of temperature. Using a Taylor series expansion for the exponent in the W.K.B. approximation for the tunnelling probability, one readily finds for the emitted current

$$j = (e^3 F^2 / 8\pi h\phi)\exp\{-4\sqrt{(2m^*)}\phi^{\frac{3}{2}}/3\hbar eF\} \qquad (3.1)$$

(cf. Mott and Sneddon 1948), where ϕ is the barrier height measured between the Fermi surface in the metal and the conduction band of the dielectric.

The simplest way to handle these modifications is to take account of the dielectric constant ϵ in the electrostatic part of the calculations and to introduce as effective electron mass m^* (appropriate to the electron in the dielectric) into the quantum-mechanical part. These two changes are incorporated in the following discussion, which follows that of Murphy and Good (1956). For the case of eqn (3.1) the potential energy external to the metal is given by $eV(x) = -eFx$, where x is

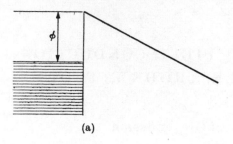

(a)

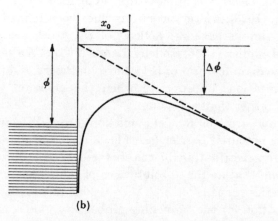

(b)

Fig. 3.1. (a) Potential-energy diagram for Fowler–Nordheim emission into a region in which the potential is due to a uniform electric field. (b) The modification in the potential-energy diagram that results from the inclusion of the image force.

the distance measured from the metal surface. If the image force is included, the potential energy becomes

$$eV(x) = -eFx - e^2/4\epsilon x \qquad (3.2)$$

(cf. Fig. 3.1(b)). The maximum value of the potential energy occurs at a point distant x_0 in front of the metal surface, where

$$x_0 = \tfrac{1}{2}(e/F\epsilon)^{\frac{1}{2}}, \qquad (3.3)$$

and the effective work function is reduced by an amount

$$\Delta\phi = (e^3 F/\epsilon)^{\frac{1}{2}}. \qquad (3.4)$$

The calculation of the field-emission current is a one-dimensional one, but the Fermi distribution of electrons inside the metal must be treated in a three-dimensional manner. It therefore becomes convenient to define a quantity

$$E_x = E - p_y^2/2m - p_z^2/2m, \qquad (3.5)$$

where E is the total energy and p_y and p_z the y and z components of momentum. In the dielectric we have

$$E_x = p_x^2/2m^* + eV(x) \qquad (3.6)$$

with $V(x)$ given by eqn (3.2), while in the metal

$$E_x = p_x^2/2m + \text{const.} \qquad (3.7)$$

The total emission current will depend on both the tunnelling probability and the number of electrons with sufficient energy incident on the metal surface. We shall consider the second factor first. The number of electrons per second per unit area incident on the surface with x-momentum in the range

$$(p_x, p_x + \mathrm{d}p_x)$$

will be given by

$$N(p_x)\,\mathrm{d}p_x = \int\!\!\int_{-\infty}^{\infty} \frac{p_x}{m} \frac{2}{h^3} \frac{\mathrm{d}p_y\,\mathrm{d}p_z\mathrm{d}p_x}{\exp\{(E - E_\mathrm{F})/k_0 T\} + 1} \qquad (3.8)$$

where E_F is the Fermi energy. Substituting for E from eqn (3.4), and writing $m\,\mathrm{d}E_x$ for $p_x\,\mathrm{d}p_x$, we find

$$N(E_x)\,\mathrm{d}E_x = \frac{4\pi m k_0 T}{h^3} \ln[1 + \exp\{(-E_x + E_\mathrm{F})/k_0 T\}]. \qquad (3.9)$$

The transmission function is found from the W.K.B. approximation to solutions of Schrödinger's equation for penetration of the barrier; it is given by

$$P(E_x) = \exp\left(-\int_{x_1}^{x_2} \left[\frac{8m^*}{\hbar^2}\{eV(x) - E_x\}\right]^{\frac{1}{2}} \mathrm{d}x\right) \qquad (3.10)$$

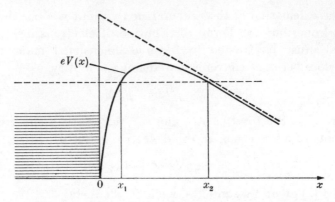

FIG. 3.2. Diagram illustrating the limits of integration in eqn (3.10) for an electron tunnelling at an energy corresponding to the horizontal dotted line.

where x_1 and x_2 are the points of zero x-component of momentum as illustrated in Fig. 3.2. Using the expression (3.2) for $V(x)$, we find

$$\ln P = - \int_{x_1}^{x_2} \left[\frac{8m^*}{\hbar^2} \left\{ -eFx + |E_x| - (e^2/4\epsilon x) \right\} \right]^{\frac{1}{2}} \mathrm{d}x, \quad (3.11)$$

where

$$x_1, x_2 = \frac{|E_x|}{2eF} \left[1 \pm \{ 1 - (e^3 F / \epsilon E_x) \}^{\frac{1}{2}} \right].$$

Burgess, Kroemer, and Houston (1953) showed that the integral could be written in terms of complete elliptic integrals of the first and second kinds; their final result is

$$\ln P = - \frac{4\sqrt{(2m^*)} |E_x|^{\frac{3}{2}}}{3\hbar eF} v(y) \quad (3.12)$$

where v is a function of the variable

$$y = (e^3 F / \epsilon)^{\frac{1}{2}} / |E_x| \quad (3.13)$$

and containing the elliptic integrals. A representative set of values of $v(y)$ is given in Table 3.1.

TABLE 3.1

Tabulation of the values of special functions used in the theory of field emission (from Good and Müller 1956).

y	$v(y)$	$t(y)$	$g(y)$
0·00	1·0000	1·0000	0·750
0·05	0·9948	1·0011	0·748
0·10	0·9817	1·0036	0·743
0·15	0·9622	1·0070	0·737
0·20	0·9370	1·0111	0·732
0·25	0·9068	1·0157	0·725
0·30	0·8718	1·0207	0·718
0·35	0·8323	1·0262	0·711
0·40	0·7888	1·0319	0·704
0·45	0·7413	1·0378	0·697
0·50	0·6900	1·0439	0·690
0·55	0·6351	1·0502	0·687
0·60	0·5768	1·0565	0·676
0·65	0·5152	1·0631	0·668
0·70	0·4504	1·0697	0·661
0·75	0·3825	1·0765	0·655
0·80	0·3117	1·0832	0·648
0·85	0·2379	1·0900	0·642
0·90	0·1613	0·0969	0·635
0·95	0·0820	0·1037	0·628
1·00	0·0000	1·1107	0·622

The number of electrons emitted per second in the energy range $(E_x, E_x + \mathrm{d}E_x)$ is given by

$$N(E_x)P(E_x)\,\mathrm{d}E_x = \frac{4\pi m k_0 T}{h^3}\ln\left\{1+\exp\left(\frac{-E_x+E_{\mathrm{F}}}{k_0 T}\right)\right\} \times$$

$$\times \exp\left\{-\frac{4\sqrt{(2m^*)}\,|E_x|^{\frac{3}{2}}}{3\hbar e F}\,v(y)\right\}\mathrm{d}E_x, \quad (3.14)$$

from eqns (3.9) and (3.12). Since the emitted electrons will have energies in the neighbourhood of the Fermi energy, we can approximate the exponent in the transmission function by the first two terms in a powers series expansion about this point,

$$\frac{4\sqrt{(2m^*)}\,|E_x|^{\frac{3}{2}}}{3\hbar eF}\, v\left\{\sqrt{\left(\frac{e^3F}{\epsilon}\right)}\Big/|E_x|\right\}$$

$$\simeq c-(E_x-E_F)/d$$

$$\simeq \frac{\sqrt{(2m^*)}\phi^{\frac{3}{2}}}{3\hbar eF}\, v\left(\frac{\Delta\phi}{\phi}\right)-(E_x-E_F)\frac{2\sqrt{(2m^*\phi)}}{\hbar eF}\, t\left(\frac{\Delta\phi}{\phi}\right) \quad (3.15)$$

where the function $t(y)$ is related to $v(y)$ by

$$t(y) = v(y)-\frac{2}{3}y\,\frac{\mathrm{d}v}{\mathrm{d}y}. \quad (3.16)$$

Values of $t(y)$ are given in Table 3.1. The total electric current is now found from the integral

$$j = e\int N(E_x)P(E_x)\,\mathrm{d}E_x \quad (3.17)$$

over the appropriate energy range. Making the Taylor series expansion of the transmission-function exponent, we have

$$j = \frac{4\pi emk_0T}{h^3}\times$$

$$\times \int \ln\left\{1+\exp\left(\frac{-E_x+E_F}{k_0T}\right)\right\}\times\exp\left[-c+\frac{E_x-E_F}{d}\right]\mathrm{d}E_x, \quad (3.18)$$

where

$$c = \frac{4\sqrt{(2m^*)}}{3\hbar eF}\, v\left(\frac{\Delta\phi}{\phi}\right) \quad (3.19)$$

and

$$d = \frac{\hbar eF}{2\sqrt{(2m^*\phi)}t(\Delta\phi/\phi)}, \quad (3.20)$$

from (3.14) and (3.15). The simplest case is the zero-temperature limit of (3.18); the appropriate range for E_x is then from $-\infty$ up to E_F, and the result is

$$j(0) = \frac{e^3Fm}{8\pi h\phi m^*}\,\frac{1}{t^2(\Delta\phi/\phi)}\,\exp\left\{-\frac{4\sqrt{(2m^*)}\phi^{\frac{3}{2}}}{3\hbar eF}\,v\left(\frac{\Delta\phi}{\phi}\right)\right\} \quad (3.21)$$

which differs from (3.1) by the multiplying functions $1/t^2(\Delta\phi/\phi)$ and $v(\Delta\phi/\phi)$ in the pre-exponential factor and the exponent respectively. Inspection of Table 3.1 shows that the factor $1/t^2(\Delta\phi/\phi)$ causes very little change in the value of $j(0)$ and it is often omitted; however, the function $v(\Delta\phi/\phi)$, occurring in the exponent, may be responsible for a change of many orders of magnitude in the calculated current density. If ϕ is expressed in eV and F in MV cm^{-1}, and if $t(\Delta\phi/\phi)$ is set equal to unity, the result for $j(0)$ in A cm^{-2} is given by

$$j(0) = 1\cdot54 \times 10^6 \frac{F^2}{\phi} \left(\frac{m}{m^*}\right) \exp\left\{-68\cdot3 \frac{\phi^{\frac{3}{2}}}{F} \left(\frac{m^*}{m}\right)^{\frac{1}{2}} v\left(0\cdot379 \frac{F^{\frac{1}{2}}}{\epsilon^{\frac{1}{2}}\phi}\right)\right\}.$$

$$(3.22)$$

In computing from this equation it is easier to regard

$$0\cdot379 F^{\frac{1}{2}}/\epsilon^{\frac{1}{2}}\phi$$

as the independent variable, and to assign to it the series of values of y listed in Table 3.1; the calculation of corresponding values of F and $j(0)$ is then a simple matter of using eqn (3.22) and Table 3.1. This procedure frequently yields only two or three points within practical limits; further points can be determined by use of eqn (3.22) with interpolated values of the functions given in Table 3.1.

At finite temperatures some electrons will have energies higher than the Fermi energy, and, since for them the tunnelling probability will be greater, an increase of current is expected with increasing temperature. The integration of eqn (3.18) for non-zero temperature is described by Good and Müller (1956) whose result is

$$j(T) = j(0) \frac{\pi k_0 T/d}{\sin(\pi k_0 T/d)},$$

$$(3.23)$$

where $j(0)$ is given by eqn (3.21) and d is given by eqn (3.20). Expressing T in degrees Kelvin, ϕ in eV, and F in MV cm^{-1}, we find

$$\frac{\pi k_0 T}{d} = 2\cdot77 \times 10^{-2} \frac{T\phi^{\frac{1}{2}}}{F}\left(\frac{m^*}{m}\right)^{\frac{1}{2}} t(0\cdot379 F^{\frac{1}{2}}/\epsilon^{\frac{1}{2}}\phi). \quad (3.24)$$

Note that the small effect of the function t cannot be ignored in this case, since the whole effect of temperature is a small one.

The validity of eqn (3.23) as an approximation clearly depends on the condition

$$k_0 T/d < 1, \qquad (3.25)$$

since if $k_0 T = d$ the right-hand side of (3.23) has a singularity that does not correspond to any physical cause. The difficulty lies in the truncation of the power series of (3.16) after two terms. The expansion of (3.16) to include an additional term yields

$$\frac{4\sqrt{(2m^*)}\,|E_x|^{\frac{3}{2}}}{3\hbar e F} v\left\{\sqrt{\left(\frac{e^3 F}{\epsilon}\right)} \Big/ |E_x|\right\} \simeq c - (E_x - E_F)/d + f(E_x - E_F)^2$$

$$= \frac{4\sqrt{(2m^*)}\phi^{\frac{3}{2}}}{3\hbar e F} v\left(\frac{\Delta\phi}{\phi}\right) - (E_x - E_F)\frac{2\sqrt{(2m^*\phi)}}{\hbar e F} t\left(\frac{\Delta\phi}{\phi}\right) +$$

$$+ (E_x - E_F)^2 \frac{2}{3}\sqrt{\left(\frac{2m^*}{\phi}\right)}\frac{1}{\hbar e F} g\left(\frac{\Delta\phi}{\phi}\right), \qquad (3.26)$$

where $g(y)$ is given by

$$g(y) = \frac{3}{4}v(y) - y\frac{\mathrm{d}v}{\mathrm{d}y} + y^2\frac{\mathrm{d}^2 v}{\mathrm{d}y^2}. \qquad (3.27)$$

Values of $g(y)$ are given in Table 3.1. It has been shown by Maserjian (1967) that an excellent approximation to the temperature dependence of the current in the region $k_0 T/d < 1$ is given by

$$j(T) = j(0)\left\{\frac{\sin(\pi k_0 T/d)}{\pi k_0 T/d} + d(f/\pi)^{\frac{1}{2}}\right\}^{-1}. \qquad (3.28)$$

This equation reduces to eqn (3.23) for most of the region $k_0 T/d < 1$ but avoids the difficulties associated with the singularity of (3.23) at $k_0 T/d = 1$. It does not of course enable extension of the calculation of the temperature dependence of the current beyond this limit.

At the zero temperature there will be no thermionic emission of electrons over the potential barrier, so that eqn (3.21) is not qualified by any conditions on the magnitude of the field strength. However, at finite temperatures thermionic emission will always dominate for sufficiently low values of the field strength, so that eqns (3.23) and (3.28) will be qualified by a condition that the field strength be sufficiently great; this condition is found by Good and Müller (1956), and can be stated approximately as

$$F \geqslant \phi^{\frac{1}{2}}T/100, \tag{3.29}$$

where F is in MV cm^{-1}, ϕ in eV, and T in degrees Kelvin. For field strengths lower than those given by eqn (3.29) thermionic emission would be expected to dominate; this will be discussed in § 3.3.

Experimental verification of this theory has been given by Lenzlinger and Snow (1969), who measured the current emission into thermally grown SiO_2 from various metal cathodes. Their results are shown in Fig. 3.3 on a Fowler–Nordheim plot,

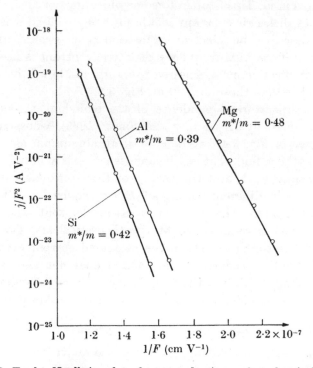

Fig. 3.3. Fowler–Nordheim plot of current density against electric field for emission from various metal electrodes into SiO_2 at room temperature. Values of effective mass are those required to fit eqn (3.21). (After Lenzlinger and Snow 1969.)

i.e. $\log(j/F^2)$ against $1/F$, for easy comparison with eqn (3.22). Reasons given for choosing SiO_2 for investigation are that photoemission work functions are well known (cf. Goodman and O'Neill 1966), and that bulk-limited effects should be less important in SiO_2 than in many other insulators. This anticipated absence of bulk-limited effects is favoured by a relatively low density of electron traps (cf. Williams 1965), and a high mobility in the conduction band (cf. Goodman 1967). The experimental points of Fig. 3.3 are fitted to eqn (3.22) with m^*/m the only adjustable parameter; the values of this quantity for best fit are shown on the diagram. The potential functions ϕ were the photoresponse thresholds of Goodman and O'Neill (1966), and for the dielectric constant the low-frequency value of 3.9 was used. This is open to some objection, since the high-frequency dielectric constant should probably be used when the model describes an electron transition over several atomic distances; if the high-frequency dielectric constant is used the values of effective mass required to fit eqn (3.21) will be about 5 % higher than those quoted on Fig. 3.3.

The temperature dependence of the emission current was also measured by Lenzlinger and Snow (1969), whose results are shown in Fig. 3.4 for the case of an aluminium electrode. These authors find that an inconsistency exists between the experimental work and the theory of this section because of the markedly different values of m^*/m required to fit eqns (3.21) and (3.23). The problem is essentially that a value of $m^*/m = 0.94$ is required to fit the theoretical singularity of eqn (3.23) to the rapid rise in the measured current occurring around 400 K. Recourse to eqn (3.28) does not help at all, since it simply removes the singularity, without changing the temperature at which the rapid rise in current takes place.

(a) Enhancement of emission by optical phonons

In the theory of field emission discussed above, the potential barrier has been calculated on the basis of classical electrostatics. In an attempt to improve on this classical continuum model of the role of the dielectric, Emtage (1967) has investigated the

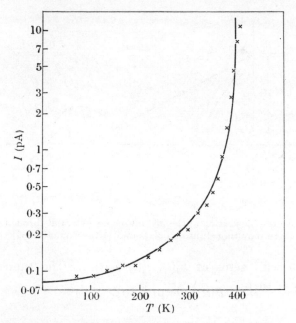

Fɪɢ. 3.4. Temperature dependence of current emission from aluminium electrodes into SiO_2 for a field of 6·1 MV cm⁻¹. The full line corresponds to eqn (3.23) with $m^*/m = 0·94$. (After Lenzlinger and Snow 1969.)

effect of interaction between longitudinal optical phonons in a polar dielectric and electrons tunnelling into it.

It is well known that there is an electric field associated with a longitudinal optical phonon in a polar solid (cf. Born and Huang 1954); the mean oscillatory field due to this cause is of order 1 MV cm⁻¹ in a normal ionic solid, which is clearly sufficient to influence the tunnelling probability. If it is assumed that the phonon frequency is much lower than the electronic frequency, then the approximation holds at each instant and the electron tunnels through a potential such as that illustrated in Fig. 3.5. In a non-polar solid, an electron from the Fermi level in the metal could tunnel from $x = 0$ to $x = x_0$ through the potential $\phi - eFx$ indicated by the solid line of Fig. 3.5. However, in a polar solid the potential will be oscillatory, and at some time t its value may be as sketched by the dotted line

4

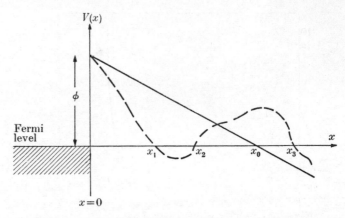

FIG. 3.5. Schematic illustration of the instantaneous potential through which an electron may tunnel into a polar solid (after Emtage 1967).

of Fig. 3.5 with zeros at $x_1(t), x_2(t), x_3(t), \ldots$ The tunnelling probability is then

$$P(t) = \exp\{-I(t)\},$$

where

$$I(t) = \frac{2\sqrt{(2m^*)}}{\hbar}\left[\left\{\int_0^{x_1(t)} + \int_{x_2(t)}^{x_3(t)} + \ldots\right\}\sqrt{\{V(x,t)\}}\,\mathrm{d}x\right] \quad (3.30)$$

where $V(x,t)$ is the oscillatory potential. Emtage (1967) bases his calculation of the emitted current on the evaluation of the integral in (3.30), and he achieved this with the help of three assumptions: that the system is one-dimensional so that all phonon wave vectors are normal to the cathode; that the integral of (3.30) is evaluated using its first term only, i.e. the term containing the integral from 0 to $x_1(t)$; and that a method of averaging over the amplitudes and phases of the phonons is adopted to obtain a fluctuating but spatially uniform average field near the cathode.

The calculations are complex but the computed results are given in dimensionless form. To this end we require the definitions of certain characteristic quantities. A quantity of the dimensions of a field strength given by

$$F_\mathrm{T} = \tfrac{4}{3}\sqrt{(2m^*)}\phi^{\frac{3}{2}}/e\hbar \quad (3.31)$$

is associated with the classical tunnelling calculation (cf. eqn (3.1)). A field strength associated with the polar phonons is given by

$$F_P = \frac{27\pi}{128} \frac{e\hbar\omega}{\epsilon^* a^2 \phi}, \tag{3.32}$$

where ω is the angular frequency of the longitudinal optical modes (supposed constant), a is the interatomic separation, and ϵ^* the effective dielectric constant defined by eqn (2.46). Also required are a characteristic current density

$$j_0 = \frac{4em^*\phi^2}{25\sqrt{(2\pi)}\hbar^3}\left(\frac{F_P}{F_T}\right)^{\frac{5}{4}}, \tag{3.33}$$

and a dimensionless number

$$A = 2(F_T/F_P)^{\frac{1}{2}} \tag{3.34}$$

With these definitions, the current density at the zero of temperature $j(0)$ is given from

$$\mathscr{J}(0) = \exp\{-Ay(\mathscr{F})\}, \tag{3.35}$$

where $\mathscr{F}$ is a dimensionless field strength defined by

$$\mathscr{F} = F/(F_P F_T)^{\frac{1}{2}} \tag{3.36}$$

and $\mathscr{J}(0)$ a dimensionless current density given by

$$\mathscr{J}(0) = j(0)/j_0. \tag{3.37}$$

The function $y(\mathscr{F})$ is given by

$$y(\mathscr{F}) = \tfrac{1}{2}\{(\mathscr{F}^{\frac{1}{2}}+z)^{-2}+z^2\}. \tag{3.38}$$

In this definition, the number z is the solution of the equation

$$z(\mathscr{F}^{\frac{1}{2}}+z)^3 = 1. \tag{3.39}$$

The results of eqn (3.35) are plotted in Fig. 3.6 both as a Fowler–Nordheim plot $j(0)/j_0$ against $1/\mathscr{F}$, and as a Schottky plot $j(0)/j_0$ against $\mathscr{F}^{\frac{1}{2}}$. Asymptotic expressions for (3.35) can be found for the limiting cases of both low and high field strengths. For high field strengths ($F > 4(F_T F_P)^{\frac{1}{2}}$), the result is

$$\mathscr{J}(0) \sim \exp(-A/2\mathscr{F}), \tag{3.40}$$

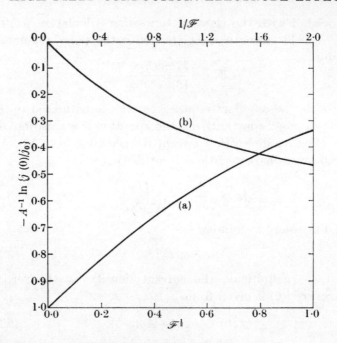

Fig. 3.6. Plot of eqn (3.35) for dimensionless current density as a function of dimensionless field strength; (a) Schottky plot, (b) Fowler–Nordheim plot. (After Emtage 1967.)

while for low field strengths $(F < (F_{\mathrm{T}}F_{\mathrm{P}})^{\frac{1}{2}}/10)$, one finds

$$\mathscr{J}(0) \sim \exp\{-A(1-\mathscr{F}^{\frac{1}{2}})\}. \tag{3.41}$$

These results can be derived analytically from eqn (3.35); they show that even in the zero-temperature approximation a low-field Schottky-type characteristic is continuous with a high-field Fowler–Nordheim characteristic.

The theory given above for the case of zero temperature includes the effects of potentials due to the zero-point motion of the ions about their equilibrium positions; increasing temperatures will affect only the field strength due to the polar phonons. In place of (3.32), we must use a temperature-dependent field strength given by

$$F_{\mathrm{P}}(T) = F_{\mathrm{P}}\coth(\hbar\omega/2k_0 T). \tag{3.42}$$

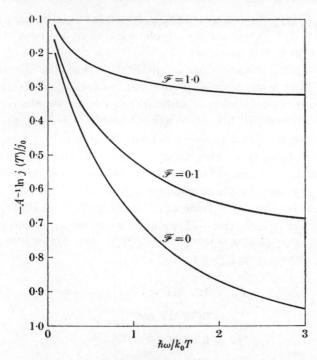

FIG. 3.7. Arrhenius plot of calculated values of temperature-dependent current for various values of the dimensionless field strength (after Emtage 1967).

Figure 3.7 shows the logarithm of the temperature-dependent current $j(T)$ as a function of the reciprocal temperature for various values of the dimensionless field strength $\mathscr{F}$. The characteristics are markedly curved; the apparent activation energy for the low-field Schottky-like emission is therefore temperature-dependent. Using eqns (3.41) and (3.42), we get for this apparent activation energy

$$\psi = \lim_{F \to 0} \left\{ -k \frac{\partial \ln j(T)}{\partial T^{-1}} \right\}$$

$$= \frac{\Lambda \hbar \omega / 4}{\tanh^{\frac{1}{2}}(\hbar \omega / 2k_0 T) \cosh^2(\hbar \omega / 2k_0 T)} . \qquad (3.43)$$

Equation (3.43) defines a quantity, ψ, which is not only temperature-dependent but bears no simple relation at all to ϕ, the work function in the usual sense.

Emtage (1967) claims only a qualitative validity for the theory of this section, and points out various quantitative defects; however a number of outstanding anomalies receive a unified treatment in his approach. Notable amongst these anomalies are the persistence of current for low field strength and temperature, the ubiquitous nature of the Schottky characteristic, and the difference between the apparent activation energy and the photoresponse threshold.

In order to illustrate orders of magnitude, consider an ionic solid for which the following parameters are assumed $\epsilon_s = 6$, $\epsilon_\infty = 3$, $\omega = 6 \times 10^{13}$ s^{-1}, $a = 3$ Å, $\phi = 1\cdot5$ eV. Then from eqns (3.31), (3.32), (3.33), and (3.34)

$$F_T = 123 \text{ MV cm}^{-1},$$

$$F_P = 0\cdot38 \text{ MV cm}^{-1},$$

$$j_0 = 5 \times 10^6 \text{ A cm}^{-2},$$

$$A = 36.$$

These values can then be used to give theoretical estimates of quantities that can be determined experimentally. From eqns (3.37) and (3.41), the current at low fields and temperatures would be
$$j(0) = j_0 \exp(-A) = 10^{-9} \text{ A cm}^{-2}.$$

The gradient of the current–voltage characteristic on a Schottky plot at low temperatures would be

$$\frac{\partial \ln \mathscr{J}(0)}{\partial F^{\frac{1}{2}}} = 2F_T^{\frac{1}{4}}/F_P^{\frac{3}{4}}$$

$$= 1\cdot4 \times 10^{-2} \text{ V}^{-\frac{1}{2}} \text{ cm}^{-\frac{1}{2}},$$

and the corresponding apparent activation energy at 300 K (from eqn (3.43)) would be $\psi = 0\cdot25$ eV. These figures are of the order of magnitude of frequently measured results.

(b) The effect of traps in the insulator

The presence of traps in the insulator may have a marked effect on the emission current. A theory to account for enhanced emission due to the presence of empty trapping centres has been given by Penley (1962) and later elaborated by Gadzuk (1970a, b). The details of the calculations are quite complex but the basic idea can be described qualitatively.

Consider an electron incident on the potential barrier whose energy is the same as that of a vacant trapping centre within the barrier. The incoming wave will grow exponentially as it approaches the trap, resonate with a large amplitude for a finite time inside the potential well of the trap, and then decay exponentially through the rest of the barrier before propagating through the conduction band of the dielectric. This process is called resonance tunnelling, and the behaviour of the wave function of the tunnelling electron shows that the transmission function will be much larger for such an electron than for an electron tunnelling by the usual process. For incident electrons with energies differing from that of the trapping level, the wave function will decay exponentially through the barrier almost as if the trap were not there. If the potential well is in the middle of the barrier the amplitudes of the waves incoming to the trap and outgoing from it can be made equal, resulting in a very large transmission function. Offsetting this large increase in the transmission function are several effects, which can be listed as follows.

(1) The supply function will be reduced because of the lifetime of the electron in the trap, since when the trap is occupied no more electrons can tunnel in this manner until it is empty again.

(2) Presumably only a certain fraction of the cathode surface area will be adjacent to trapping centres; the total emission current will be the sum of the resonance enhanced contribution from these areas and simple field emission from the remainder.

(3) As mentioned above, the electron energy must be the

same as the energy of the trapping centre; not all electrons coming from the cathode and adjacent to an empty trapping centre have their transmission function enhanced as described.

The total current emitted through a barrier containing trapping centres could be strongly temperature-dependent for a variety of reasons, as follows.

(1) Electrons may be thermally ionized from trapping levels, so that increasing temperature may lead to the opening of a larger number of channels for resonance tunnelling.

(2) If the energy of the trapping levels is slightly above the Fermi energy, increasing temperature will result in a larger fraction of incident electrons having the appropriate energy for resonance tunnelling.

(3) Increasing temperature will broaden the trapping level, which also results in a larger fraction of incident electrons with energies in a range suitable for resonance tunnelling.

(4) The empty trapping levels in question may be highly immobile, positively charged ions; with increasing temperature their mobility may be sufficiently increased to enhance their density adjacent to the cathode. This would increase the number of channels available for resonance tunnelling.

There does not seem to be any direct experimental evidence of the existence of resonance tunnelling into dielectrics. However, Plummer and Young (1970) have used this theory to interpret the results of experiments on field emission from a tungsten cathode into vacuum with alkaline-earth contaminants on the cathode.

3.2. Tunnelling through thin insulating films

If an insulating layer is so thin that an electron may tunnel directly from cathode to anode without at any time occupying the conduction band of the dielectric, the theories given above require substantial modification. The film thickness for which this will occur depends on both the work function and the applied voltage; generally speaking, a thickness $\sim$30 Å or less will involve direct tunnelling between the electrodes for typical experimental conditions. The theory of this process is extensive and technically important, and is reviewed at length by Duke (1969); it is discussed briefly below since it is of only peripheral importance to the question of current conduction and breakdown.

Frenkel (1930) gave an approximate analysis of electron tunnelling through a thin insulating film using a simple rectangular potential barrier model, and a zero-temperature approximation. Extensions to this work and modifications of it were derived by Sommerfeld and Bethe (1933), Holm (1951), and Simmons (1963a, b, 1964). However, the most frequently quoted treatment is that of Stratton (1962), which is outlined below.

Stratton's method follows closely the method of Murphy and Good (1956) for the calculation of emission into thick dielectric; we describe the results in terms of modifications to the theory given above. The Fowler–Nordheim emission current is determined by eqns (3.23) and (3.21) together with the definitions of the auxiliary functions v and t (given numerically in Table 3.1) and the definitions of c and d (eqns (3.19) and (3.20)). The thin-film tunnelling current is given by similar formulae, subject to two modifications.

(1) The functions v and t (and therefore also c and d) are defined in terms of the tunnelling integral (3.10) whose integrand is a function of the potential barrier taken between the classical turning points x_1 and x_2, as illustrated in Fig. 3.2. Because the anode is immediately adjacent in the thin-film case, the image force due to it will modify the potential barrier. This is illustrated in Fig. 3.8; both the integrand and the limits of integration will be changed, and the special functions to be used will therefore be analogous to v, t, c, and d.

(2) In calculating the zero-temperature emission current given by (3.21), the integration (3.18) was taken between the Fermi level in the cathode and $-\infty$. For the thin-film case the lower limit $-\infty$ must be replaced by $-eV$, since an electron may not tunnel into the anode below its Fermi level. This

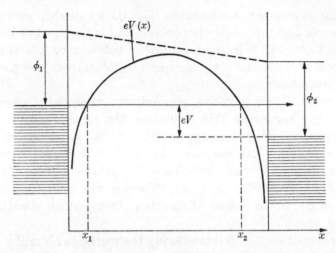

Fig. 3.8. Diagram of the potential model used by Stratton (1962) for the calculation of the tunnelling current between two electrodes with potential difference V.

change of limits results in an additional factor in the expression for the current; it is of order zero for low voltages across the film, and of order unity for high voltages.

If current tunnelling in the reverse direction is neglected in a first approximation, Stratton's (1962) final results corresponding to eqns (3.23) and (3.21) are

$$j(0) = \frac{4\pi m e}{c_1^2 h^3} \exp(-b_1)\{1 - \exp(-c_1 e V)\} \qquad (3.44)$$

and

$$j(T) = j(0)\pi c_1 k_0 T / \sin(\pi c_1 k_0 T). \qquad (3.45)$$

There has been an unfortunate difference in the notation for the auxiliary functions of Murphy and Good (1956) and Stratton (1962). The correspondence is given in Table 3.2, from which it is clear that eqns (3.44) and (3.45) correspond very closely to eqns (3.21) and (3.23).

Stratton's results are quite general within the limitation of the independent-electron model and do not depend on the potential $V(x)$ of Fig. 3.8 being a trapezoidal barrier modified by image forces. In fact, explicit calculations are given of the

TABLE 3.2

*Correspondence between the auxiliary functions
defined in theories of field emission and thin-film
tunnelling.*

Auxiliary function defined by Murphy and Good (1956)	Analogous function defined by Stratton (1962)
c	b_1
d	$1/c_1$

auxiliary functions b_1 and c_1 for various assumed tunnelling potentials. In addition, the theory is further developed in a general way for small applied voltages, without specification of the exact barrier shape. To this end, b_1 and c_1 are expanded as quadratic power series in the applied potential,

$$b_1 = b_{10} - b_{11}(eV) + b_{12}(eV)^2, \qquad (3.46)$$

$$c_1 = c_{10} - c_{11}(eV) + c_{12}(eV)^2. \qquad (3.47)$$

Substitution in the equations (3.44) and (3.45) then yields after suitable approximations

$$j(0) = \frac{4\pi me}{c_{10}^2 h^3} \exp(-b_{10})\{1 - \exp(-c_{10}eV)\}\exp(-b_{11}eV - b_{12}e^2V^2)$$

$$(3.48)$$

and

$$j(T) = j(0)\pi c_{10}k_0 T / \sin(\pi c_{10}k_0 T). \qquad (3.49)$$

The zero-temperature current density of eqn (3.48) can be expanded as a power series

$$j(0) = \frac{4\pi me^2 V}{c_{10}h^3} \exp(-b_{10}) \times$$

$$\times \{1 + (b_{11} - \tfrac{1}{2}c_{10})eV(\tfrac{1}{2}b_{11}^2 - \tfrac{1}{2}b_{11}c_{10} + \tfrac{1}{6}c_{10}^2 - b_{12})e^2V^2 + \ldots\}.$$

$$(3.50)$$

Stratton (1962) shows that for the case of a symmetrical barrier $b_{11} = c_{10}/2$; all terms even in the applied voltage then

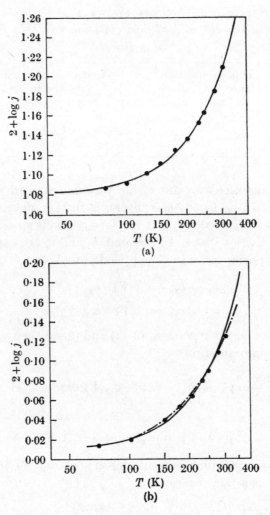

FIG. 3.9. The dependence of the current density on the temperature for Al$_2$O$_3$ films ~30 Å thick at (a) 0·5 V and (b) 0·05 V. The points are experimental data, and the curves correspond to eqn 3.45. (After Hartman and Chivian 1964.)

vanish from (3.50) to yield the result

$$j(0) = \frac{4\pi m e}{c_{10} h^3} \exp(-b_{10}) \left\{ eV + \left(\frac{c_{10}^2}{24} - b_{12} \right) (eV)^3 + \ldots \right\} \quad (3.51)$$

Experimental results on various substances have been fitted to Stratton's theory. Tunnelling in thin films of Al_2O_3 has been investigated by Fisher and Giaever (1961), Meyerhofer and Ochs (1963), and Hartman and Chivian (1964); the last authors have presented their data in such a way as to permit direct comparison with Stratton's theoretical results. Figure 3.9 shows the temperature dependence of the current density for fixed values of the voltage applied to Al_2O_3 films $\sim$30 Å thick. With the use of eqn (3.45) these results can be used to determine the auxiliary function c_1 as a function of the applied voltage. With $c_1(V)$ known, $b_1(V)$ can be calculated from eqn (3.44), and also the extrapolated zero-temperature value of the current. The results of this process are shown as the experimental points in Fig. 3.10; also shown on this diagram are the parabolas obtained by fitting eqns (3.46) and (3.47) to the lower-voltage parts of the experimental curves. The coefficients in the power-series expansions obtained in this way are shown in Table 3.3. It will be observed that, although the low-voltage expression for c_1 is approximately correct to substantially higher voltages than those for which it was fitted, the expression for b_1 becomes grossly inaccurate outside its region of fit. The critical role of b_1 in determining the magnitude of the current means that the low-voltage approximation must be treated strictly as such. Hartman and Chivian (1964) also made a Fowler–Nordheim plot of the current–voltage characteristic for a thick film of Al_2O_3, and estimated the metal–insulator work function to be 0·85 eV.

Extensive measurements of the current–voltage characteristics of mica films have been made by McColl and Mead (1965). For the case of very thin films (30 and 40 Å) they also interpret their results in terms of Stratton's theory, but in a slightly different way from that quoted above. Data for green Muscovite mica 40 Å thick with one aluminium and one gold electrode are

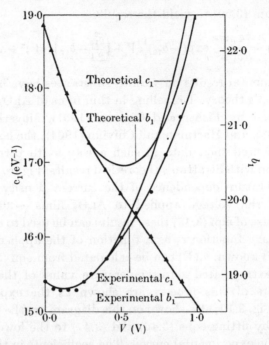

FIG. 3.10. The dependence of the coefficients b_1 and c_1 on voltage. The curves labelled 'experimental' are found from experimental data using eqns (3.44) and (3.45) as explained in the text; the curves labelled 'theoretical' result from a fit of eqns (3.46) and (3.47) to the part of the experimental curves lying in the region $0 < V < 0.3$ V. (After Hartman and Chivian 1964.)

TABLE 3.3

Coefficients for evaluation of Stratton's auxiliary functions b_1 and c_1 in the low-voltage limit for Al_2O_3 films (after Hartman and Chivian 1964).

Coefficients in series (3.46) for b_1	Coefficients in series (3.47) for c_2
$b_{10} = 22.37$	$c_{10} = 15.38 \ (eV)^{-1}$
$b_{11} = 7.56 \ (eV)^{-1}$	$c_{11} = 1.32 \ (eV)^{-2}$
$b_{12} = 7.4 \ (eV)^{-2}$	$c_{12} = 5.29 \ (eV)^{-3}$

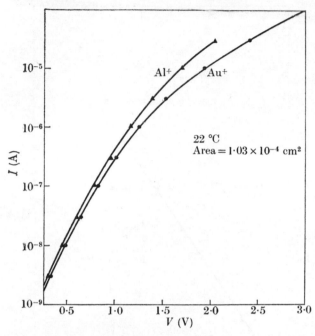

FIG. 3.11. Current–voltage characteristics of a 40 Å green Muscovite mica film with one aluminium and one gold electrode (after McColl and Mead 1965).

shown in Fig. 3.11; since the effect of electrode polarity is very slight, comparison was made with the theoretical results (3.49) and (3.51). At very low voltages, the linear term in eqn (3.51) is dominant and the film is ohmic; if the ohmic current is subtracted from the total current, then the remaining current should be cubic in the applied voltage. For the film in question the ohmic resistance was $2 \cdot 31 \times 10^8 \, \Omega$, and the results of carrying out the procedure indicated are shown in Fig. 3.12. The coefficients in the expansion of b_1 and c_1 resulting from this analysis are shown in Table 3.4. Only the constant term in the series for c_1 is found by this method; to find the additional coefficients one would require sets of data such as those shown in Fig. 3.9 for Al_2O_3, and these were not given for mica by McColl and Mead. Using details of the theory of the image force barrier, they quoted a metal–insulator work function of

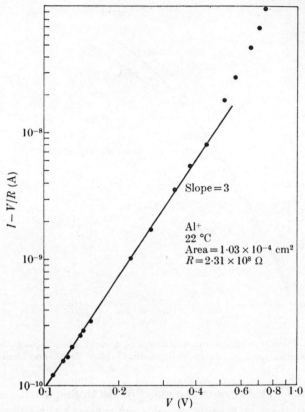

FIG. 3.12. Cubic dependence of $I - V/R$ on voltage up to approximately 0·45 V for a 40 Å green Muscovite film (after McColl and Mead 1965).

TABLE 3.4

Coefficients for evaluation of Stratton's auxiliary functions b_1 and c_1 in the low-voltage limit for green Muscovite mica films (after McColl and Mead 1965).

Coefficients in series (3.46) for b_1	Coefficients in series (3.47) for c_1
$b_{10} = 31·1$	$c_{10} = 22·5 \ (\text{eV})^{-1}$
$b_{11} = 11·2 \ (\text{eV})^{-1}$	c_{11} $\Big\}$ Not evaluated
$b_{12} \simeq 0$	c_{12}

0·95 eV for gold–mica, in conjunction with an effective electron mass of $m^* = 0.92m$. Photoresponse measurements gave a threshold of 0·8 eV, which is in reasonable agreement with the figure from tunnelling experiments.

Various difficulties are noted by McColl and Mead (1965) with regard to the universality of these figures; they were unable to reproduce identical results when the aluminium electrode was negative, or when the mica film was 30 Å thick. However, they conclude that the basic nature of the mechanism is established.

Current–voltage characteristics for thicker films were also measured by McColl and Mead (1965) to ascertain whether Fowler–Nordheim emission into conduction levels was the operative process; their results are shown in Fig. 3.13. Three reasons are given for the inapplicability of the theory of § 3.1 to this case. First, there is a marked thickness dependence of the current. McColl and Mead note that this can be removed if the voltage across the samples is written as $(V+2)$ rather than V; the origin of this 2-volt bias is, however, not explained. Secondly, the characteristics of Fig. 3.13 are distinctly curved over their whole length. This circumstance is well explained by the Emtage theory of phonon-assisted emission given in § 3.1(a). It is also noted by Emtage (1967) that the phonon-assisted emission mechanism can explain at least part of the thickness dependence referred to above. Finally, the magnitude of the current is lower by a factor $\sim 10^9$ than that predicted by eqn (3.21) with $\phi = 0.95$ eV, as determined from thin-film tunnelling. Equation (3.40) does in fact lead to a lower value of current density than that given by eqn (3.21), but only by a factor of about 10^3 for the case in hand. McColl and Mead (1965) advance an alternative explanation. It is argued that the conduction band of the insulator is relatively narrow, so that the wave functions for conduction electrons are highly localized. Such localized states will be nearly orthogonal to the plane-wave states of the metal, so that transitions that would be allowed by conservation of energy alone may nevertheless be improbable, owing to the near orthogonality of the initial and final states.

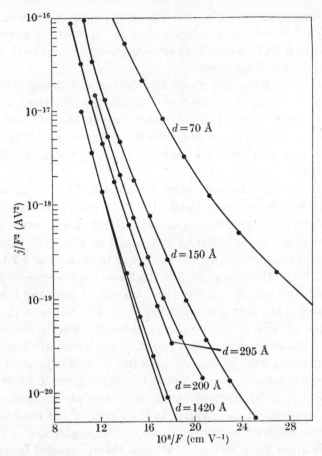

FIG. 3.13. Fowler–Nordheim plots of the conduction in green Muscovite mica films of various thicknesses at 77·4 K (after McColl and Mead 1965).

A further fundamental development of tunnelling has been made by Maserjian (1967), who has investigated the theory of the case in which the potential barrier is concave upwards rather than convex as in Fig. 3.8; such a situation could arise in the presence of ionic space charge. The results of this work have been applied to tunnelling in thin films of TiO_2 by Maserjian and Mead (1967); a similar explanation has been advanced by Gundlach (1969) to explain tunnelling in Al_2O_3 films.

3.3. Schottky emission

Schottky emission is essentially thermionic emission from a metal electrode into the conduction band of a dielectric, with the image force correction taken into account. The appropriate potential barrier is shown in Fig. 3.1(b), and the emission current is

$$j = AT^2 \exp\{-(\phi - \Delta\phi)/k_0 T\}$$
$$= AT^2 \exp\left\{\frac{-(\phi - e^{\frac{3}{2}} F^{\frac{1}{2}}/\epsilon^{\frac{1}{2}})}{k_0 T}\right\}. \tag{3.52}$$

The lowering of the barrier, $\Delta\phi$, has been obtained from eqn (3.4). The constant A is the familiar Richardson–Dushman constant of thermionic emission, given by

$$A = 4\pi emk_0^2/h^3 \simeq 120 \text{ A cm}^{-2} \text{ K}^{-2}. \tag{3.53}$$

It is clear from eqn (3.52) that the logarithm of the current is a linear function of the square root of the field strength; results plotted with the axes marked in this way are referred to as a Schottky plot, as mentioned in § 3.1(b). Pollack (1963) measured the current through thin films of Al_2O_3 with Pb electrodes, and his results are shown as a Schottky plot in Fig. 3.14. In order to verify the temperature dependence predicted by eqn (3.52), the logarithm of the current was plotted as a function of temperature as shown in Fig. 3.15; for temperatures lower than 235 K the current was approximately temperature-independent and was ascribed to tunnelling. The Schottky current is therefore the total current minus the tunnelling current, and this is displayed in Fig. 3.16 in a form suitable for comparison with eqn (3.52). Pollack assumed a dielectric constant of 12 and using (3.52) found for the work function $\phi = 0.64$ eV. This value is lower than might have been expected, and is attributed to the formation of a positive space-charge layer in the oxide adjacent to the cathode.

It has been mentioned above that Fisher and Giaever (1961) found tunnelling in Al_2O_3 films ~50 Å thick over a wide range of temperature; in the work just cited Pollack found Schottky emission in a film 340 Å thick above 235 K. More recently

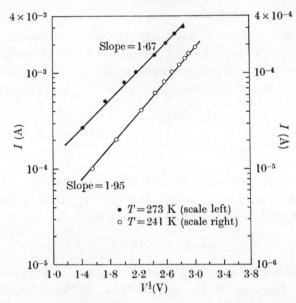

FIG. 3.14. Schottky plot of current–voltage characteristics for a Pb–Al₂O₃–Pb sandwich with a 340 Å thick insulating film (after Pollack 1963).

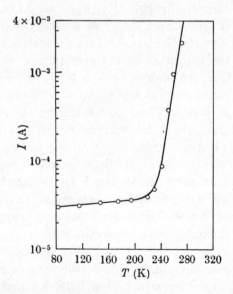

FIG. 3.15. Temperature dependence of the current for a Pb–Al₂O₃–Pb sandwich with 7 V applied to the 340 Å insulating film (after Pollack 1963).

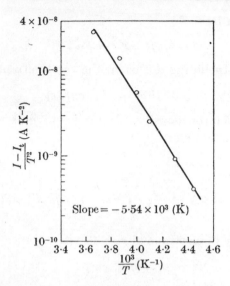

FIG. 3.16. The logarithm of the Schottky current as a function of the reciprocal of the temperature for a Pb–Al$_2$O$_3$–Pb sandwich with 7 V applied to the 340 Å insulating film (after Pollack 1963).

Antula (1971) has found Schottky emission at room temperature in films ~100 Å thick, both pure Al$_2$O$_3$ and doped with positive metallic ions. Experimental work which has been interpreted as Schottky emission is by no means confined to these several examples; the results above are fairly typical.

(a) The effect of diffusion

It is tacitly assumed in the derivation of eqn (3.52) that all of the current thermally emitted by the cathode is carried away in the conduction levels of the insulator. If the mobility in the insulator is sufficiently low, it will control the total current and a diffusion-limited theory will be required. Such theories are well known for Schottky barriers in metal–semiconductor junctions (cf. Landsberg 1951), and a modification for the case of a metal–dielectric interface has been given by Emtage and O'Dwyer (1966).

If image forces are taken into account, the potential in the dielectric is given by

$$V(x) = \phi/e - Fx - e/4\epsilon x \qquad (3.54)$$

(cf. Fig. 3.1(b)). Including diffusion, the total current density can be written

$$j = ne\mu(\partial V/\partial x) + k_0 T\mu(\partial n/\partial x) \qquad (3.55)$$

where n is the electron density. Taking j and μ as constants and solving for n, we find

$$n(x) = \exp\{-eV(x)/k_0 T\} \times$$

$$\times \left[N + \frac{j}{k_0 T\mu} \int_0^x \exp\{eV(x)/k_0 T\} \, dx \right], \quad (3.56)$$

where the density of electrons at $x = 0$, viz.

$$N = 2(2\pi m k_0 T/h^2)^{\frac{3}{2}}, \qquad (3.57)$$

has been used as a boundary condition.

To calculate the integral of eqn (3.56) for the potential function (3.54), we require the following results:

$$\int_0^\infty \exp\{-(ax+b/x)\} \, dx \simeq 1/a \quad \text{if} \quad (ab)^{\frac{1}{2}} \ll 1 \qquad (3.58)$$

$$\simeq \pi^{\frac{1}{2}}(b/a^3)^{\frac{1}{4}} \exp\{-2(ab)^{\frac{1}{2}}\} \quad \text{if} \quad (ab)^{\frac{1}{2}} \gg 1, \quad (3.59)$$

and

$$\int_0^x \exp\{-(ax+b/x)\} \, dx$$

$$\simeq \int_0^\infty \exp\{-(ax+b/x)\} \, dx - \frac{1}{a} \exp(-ax) \quad \text{if} \quad x \gg (b/a)^{\frac{1}{2}}.$$

$$(3.60)$$

Using (3.58) and (3.60) in (3.56), we find for the low-field approximation

$$n(x) = j/e\mu F + \exp(eFx/k_0 T) \times \{N \exp(-\phi/k_0 T) - j/e\mu F\}.$$

To avoid divergence of the particle density the coefficient of the exponential term must vanish, so that

$$j = Ne\mu F \exp(-\phi/k_0 T), \tag{3.61}$$

which is the usual result for low fields. For the high-field case we use (3.59) and (3.60) in (3.56) to yield for large values of x

$$n(x) = j/e\mu F + \exp(eFx/k_0 T) \times$$

$$\times \left\{ N \exp(\phi/k_0 T) + \frac{j}{k_0 T \mu} \frac{(\pi k_0 T)^{\frac{1}{2}}}{(4eF^3\epsilon)^{\frac{1}{4}}} \exp(-\Delta\phi/k_0 T) \right\},$$

from which we obtain

$$j = Ne\mu \left(\frac{k_0 T}{\pi}\right)^{\frac{1}{2}} \left\{ 4\epsilon \left(\frac{F}{e}\right)^3 \right\}^{\frac{1}{4}} \exp\{-(\phi - \Delta\phi)/k_0 T\}. \tag{3.62}$$

Equation (3.62) gives the current density when the mobility is so low that the current is diffusion-controlled. The ratio of the diffusion-controlled current (3.62) to the Schottky current given by (3.52) and (3.53) is given by

$$\frac{j(3\cdot 62)}{j(3\cdot 52)} = \mu\sqrt{(2m)}\{4\epsilon(F/e)^3\}^{\frac{1}{4}}. \tag{3.63}$$

If this ratio is less than unity the current will be diffusion-controlled; putting typical values in (3.63), we see that this will occur for electron mobilities in the insulator given by

$$\mu \lesssim 5/F^{\frac{3}{4}}, \tag{3.64}$$

where μ is in $cm^2\ V^{-1}\ s^{-1}$ and F in MV cm^{-1}. There does not seem to be any clear evidence of diffusion-limited Schottky emission. Certainly the experimentally measured pre-exponential factor is frequently orders of magnitude lower than that given by eqn (3.53), but there are various possible expanations for this effect. In addition to the one already described quantitatively by eqn (3.63), it seems likely that emission from the cathode takes place at a number of preferred locations, which together total only a small fraction of the cathode surface area.

It has been assumed in the above treatment that the space charge is not sufficiently intense to cause an appreciable correction term in the potential (3.54); this point will be taken up again in a later chapter.

HIGH FIELD CONDUCTION:
BULK EFFECTS

4.1. High field ionic conduction

IN describing a conduction process as a bulk process from a theoretical point of view one is seeking to determine the field dependence of the conductivity (1.1), on the assumption that the field itself is uniform and that the flow of charge carriers is determined only by the properties of the dielectric. In this connection, it should be noted that, although one depends on the electrodes for continual supply of charge carriers, a process is not regarded as a bulk process if electrode emission changes the charge density within the dielectric. On the assumption that we are dealing with a single species of charge carrier, the investigation is therefore confined to field dependence of the mobility or the charge-carrier density; the latter may of course be field-dependent even though the charge density is not, since field-enhanced ionization will increase the number of charge carriers without changing the charge density.

(a) Field-dependent ionic moblity

A field-dependent ionic mobility was first calculated by Mott and Gurney (1948). For this purpose they used eqn (2.3) to determine the mobility without invoking a low-field approximation. The result is readily found to be

$$\mu = (2a\nu_0/F)\exp(-\Delta g/k_0 T)\sinh(eFa/2k_0 T). \qquad (4.1)$$

If $eFa \ll k_0 T$ (as is frequently the case), the sinh term in (4.1) can be expanded to yield

$$\mu(F) \simeq \mu(0)\{1 + \tfrac{1}{3}(eFa/2k_0 T)^2\} \qquad (4.2)$$

where $\mu(0)$ is the low-field mobility.

Several attempts have been made to apply eqn (4.1) to experimental data, but the outcome has not been entirely

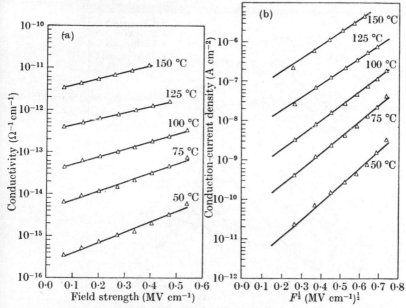

FIG. 4.1. The high-field conduction properties of KCl for various temperatures; (a) the logarithm of the conductivity as a function of the field strength; (b) the logarithm of the current density as a function of the square root of the field strength. In both cases the data extrapolated to zero field yield straight lines on an Arrhenius plot. For the conductivity the apparent activation energy is 1·08 eV, while for the current density the value is 1·18 eV. (After Hanscomb *et al.* 1966.)

successful. Hanscomb, Kao, Calderwood, O'Dwyer, and Emtage (1966) found that the conductivity of NaCl and KCl was appreciably non-ohmic for fields less than 1 MV cm⁻¹ at room temperature; their results for KCl are shown in Fig. 4.1. Since the jump distance a is accurately known for these substances, there are no disposable parameters in the field-dependent term of (4.2) which is very small for the conditions of the experiment; field-dependent ionic mobility was therefore rejected as a possible explanation of the non-ohmic behaviour of NaCl and KCl. Alternative explanations for these data will be examined below.

High-field conduction of various glasses was investigated by Vermeer (1956a), whose results are shown in Fig. 4.2. Analysis

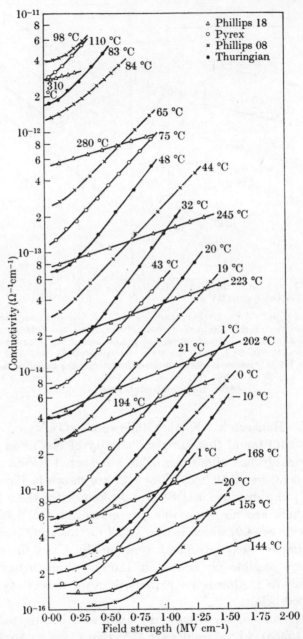

Fig. 4.2. The effect of field strength on the conductivity of different glasses at various temperatures. The Na$_2$O content of the glasses listed was Phillips 18, very low; Pyrex, 3·5 %; Phillips 08, 5·1 %; Thuringian, 12·8 %. (After Vermeer 1956a.)

TABLE 4.1

Constants characteristic of the high-field con-
ductivity of various glasses as interpreted in terms
of equation (4.1) (after Vermeer 1956a).

Glass	Δg (eV)	a (Å)
Phillips 18	1·43	7·0
Pyrex	0·96	14·0
Phillips 08	0·91	13·0
Thuringian	0·96	15·5

of these data in terms of eqn (4.1) yielded the parameter values
given in Table 4.1. Vermeer (1956a) regarded these constants
as phenomenological data only, possibly because the ionic jump
distances range up to about five times the interatomic distance.
It is indeed difficult to reconcile such large jump distances with
the assumptions made in the derivation of eqn (4.1); in this
connection, it should be noted that a smooth potential barrier
such as that sketched in Fig. 2.3 is not necessary for the
validity of (4.1), but it is necessary that the potential barrier
have its maximum midway between the equilibrium positions.

Polyvinyl chloride has been investigated by Ieda, Kosaki,
and Sugiyama (1970); their high-field conduction results are
shown in Fig. 4.3, which also illustrates the fit to eqn (4.1).
This fit to the equation of field-dependent mobility requires
that the ionic jump distance be a function of both temperature
and degree of plasticity, as shown in Fig. 4.4. The large change
in the jump distance occurs in the transition region between the
glassy and rubbery states of the polymer. Ieda *et al.* (1970)
have also considered the effect of introducing the Onsager
effective field rather than the macroscopic field in eqn (4.1).
The related questions concerning the introduction of either
effective charges or effective field strengths in the hyperbolic-
sine term of eqn (4.1) have been treated in detail by Dignam
(1968); he concludes that even at very high field strengths,
where the net activation energy for ion migration must be

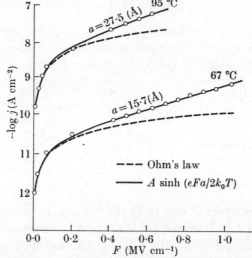

FIG. 4.3. Current density as a function of field strength for unplasticized polyvinyl chloride below and above the glass transition temperature (after Ieda *et al.* 1970).

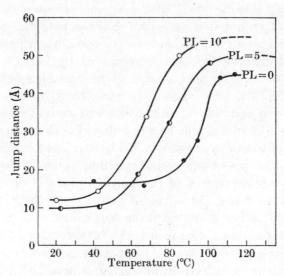

FIG. 4.4. Temperature dependence of apparent ionic jump distances for unplasticized (PL = 0) and plasticized (PL = 5 and PL = 10) polyvinyl chloride (after Ieda *et al.* 1970).

represented as a power series in the field strength, the macroscopic Maxwell field F and the charge e carried by the defect are the appropriate quantities to use in the activation-energy term linear in the field strength, provided that the potential barrier is symmetrical. No correction for effective charge or effective field strength is appropriate even in the presence of partial covalent bonding or polarization phenomena.

Over-all, it seems that the field dependence of the mobility predicted by eqn (4.1) is too small to account for the observations of non-ohmic conduction.

(b) Field-dependent carrier density

The possibility of a field dependence of the charge-carrier density for ionic solids arises only for the case of extrinsic conductivity, and then only if conditions are such that the degree of dissociation of divalent impurities and vacancies is small. In the presence of an applied field, the Gibbs free energy of association of a divalent impurity and a vancancy is no longer the same for all twelve nearest-neighbour vacancy sites. Consider a divalent cationic impurity associated with a cation vacancy; Fig. 2.4 can be used to illustrate the situation, with the shaded square now representing the divalent impurity and the shaded circles the possible vacancy positions. If the electric field is in the 100 direction, it will not change the energy of association for the four transverse vacancy positions, and will cause a change $\pm eFa$ for the two remaining groups of four positions. If these considerations are used to modify eqn (2.31), an additional field-dependent term arises of the same form and magnitude as that in eqn (4.2); it is therefore inadequate to explain the much larger experimental deviations from ohmic behaviour. The source of the difficulty is that in both cases the product of the ionic charge and the potential drop of the applied field over interionic distances is less than the thermal energy.

Hanscomb et al. (1966) made an order-of-magnitude calculation in an attempt to overcome this difficulty. What is required is some criterion to distinguish between cation vacancies that are associated with the divalent cation impurity' and those that are not. This problem has been discussed by Lidiard (1957), and it seems to have no simple answer. In a purely arbitrary assumption, we can consider the vacancy to be associated if its energy change in moving one hopping distance towards the divalent impurity is at least equal to $k_0 T$. The radius within which the complex is associated will then be given by r_0, where

$$ea\left[\frac{\partial}{\partial r}\left(\frac{e}{\epsilon_s r}\right)\right]_{r=r_0} = -k_0 T, \quad \text{i.e.} \quad r_0^2 = e^2 a/\epsilon_s k_0 T. \quad (4.3)$$

Substituting values for KCl in eqn (4.3) one finds that this convention leads to a radius of approximately 20 Å.

The energy corresponding to the product of an ionic charge and the potential drop of the external field over a distance of 20 Å may easily exceed the thermal energy; significant non-ohmic behaviour should therefore occur in a model which regards the complex as associated up to a distance given by eqn (4.3). Since the detailed statistics would be somewhat complex, we adopt a simple continuum model to investigate the conductivity. Consider the cation vacancy to be trapped by the divalent impurity in an otherwise structureless dielectric. Under the influence of an applied field, the ionization energy will be modified by a Schottky term $2(e^3 F/\epsilon_s)^{\frac{1}{2}}$, which is greater than that given by eqn (3.4) by a factor 2, since in this case the attraction is between two charges rather than between a charge and its image. The number of cation vacancies ionized per second per unit volume will then be given by

$$n = n_I \nu \exp[\{-\zeta + 2(e^3 F/\epsilon_s)^{\frac{1}{2}}\}/k_0 T], \quad (4.4)$$

where ν is a frequency, of order of the frequency of hops between adjacent vacancy positions, n_I the density of divalent cation impurities, and ζ the energy of association of the complex. In the steady state, eqn (4.4) will also give the rate of recombination of cation vacancies with divalent impurities per unit volume. Denoting the mean free path of a cation vacancy by L, we have as an estimate to the mean free time

$$\tau = L/\mu_{CV} F. \quad (4.5)$$

This assumes that the carrier velocity is solely determined by the applied field; the higher the field the more nearly correct will this assumption be, while for low field strengths random diffusion velocity will dominate. From eqns (4.4) and (4.5), the mean density of dissociated cation vacancies will be given by

$$n_{CV} = \frac{n_I \nu L}{F} \exp[\{-\zeta + 2(e^3 F/\epsilon_s)^{\frac{1}{2}}\}/k_0 T]. \qquad (4.6)$$

The order of magnitude of the free path is estimated as being the length of a cylinder whose radius is given by (4.3) and which contains one ionized trap, i.e.

$$\pi r_0^2 L = 1/n_{CV},$$

or, from (4.3),
$$L = (\epsilon_s k_0 T/n_{CV} \pi e^2 a). \qquad (4.7)$$

Substituting (4.7) in (4.6) and solving for n_{CV}, one obtains

$$n_{CV} = (n_I \nu \epsilon_s k_0 T/\pi e^2 a \mu_{CV} F)^{\frac{1}{2}} \exp[\{-\zeta + 2(e^3 F/\epsilon_s)^{\frac{1}{2}}\}/2k_0 T]. \qquad (4.8)$$

In order to collect all the exponentially temperature-dependent terms in (4.8) we note from eqn (2.6) that

$$\mu_{CV} = (4a^2 e \nu_0/k_0 T) \exp(-\Delta g_{CV}/k_0 T). \qquad (4.9)$$

To estimate the temperature dependence of ν we write

$$\nu = (k_0 T/h) \exp(-\Delta g_{CV}/k_0 T), \qquad (4.10)$$

assuming that the same free energy of activation is appropriate. Using eqns (4.8), (4.9), and (4.10), we get for the current density

$$j = (4n_I a e \nu_0 k_0 T \epsilon_s F/\pi h)^{\frac{1}{2}} \exp[\{-\Delta g_{CV} - \zeta/2 + (e^3 F/\epsilon_s)^{\frac{1}{2}}\}/k_0 T]. \qquad (4.11)$$

Comparison of experimental data with eqn (4.1) was given by Hanscomb et al. (1966); the apparent activation energy deduced from the Schottky plots of Fig. 4.1 was 1·18 eV. This should be compared with the activation enthalpy for the motion of a cation vacancy plus one half of the enthalpy of association of the divalent impurity and the cation vacancy.

Use of Tables 2.1 and 2.2 gives an average value of 0·94 for this quantity, showing only fair agreement with the result from high-field data. The slope of the lines in Fig. 4.1(b) yields a value of 1·9 for the dielectric constant, which is low by a factor of two, since the static dielectric constant should be used in this case.

If, on the other hand, one uses the straightforward Schottky emission equation (3.52) for comparison with the experimental results of Fig. 4.1(b), a dielectric constant of 1·9 is in good agreement with theory, since the high-frequency value is appropriate in this case. The value 1·18 eV would then correspond to the cathode–dielectric work function, and is much lower than might have been expected.

4.2. High-field electronic mobility

Just as for ionic conduction, the effect of a high field on the bulk electronic conductivity of insulators may be either to change the mean mobility per carrier, or to increase the carrier density. In this section we give a short discussion of the first of these topics, and reserve the question of carrier multiplication for the following section. The field dependence of the mean mobility in semiconductors has been extensively reviewed by Conwell (1967), and applied particularly to n-type germanium, for which there exists a great deal of information concerning electron conduction band structure and lattice vibrational modes. Such information is simply not available for insulators, and it is therefore not appropriate to consider any but the simplest models of conduction.

In the band theory of conduction, the relaxation time $\tau(E)$ for an electron of energy E interacting with lattice vibrations at temperature T is defined and calculated explicitly in §2.2 for the case of interaction with longitudinal optical modes of vibration of a polar crystal (eqn (2.55)) and longitudinal acoustic modes of a non-polar crystal (eqn (2.61)). There is nothing in these derivations to cause the field strength to appear explicitly in the mobility, provided that the phonon distribution remains in thermal equilibrium. The field therefore influences the mobility of the carriers by increasing their energy; it is clear from eqns (2.55), (2.61), and (2.63) that one would expect an increase in the mobility for polar crystals and a decrease for non-polar crystals.

This effect of the field in increasing the mean energy of the electrons is referred to as heating, and it has become customary to refer to 'hot electrons' whether or not a temperature can be assigned to them in the sense explained in §2.4(a). It is clear that a calculation of the field-dependent mobility requires a knowledge of the energy distribution function of the conduction electrons. This in turn requires solution of a Boltzmann equation containing terms describing:

(1) Acceleration due to the applied field.

(2) Collisions with other conduction electrons.

(3) Collisions with lattice vibrations.

(4) Ionization of electrons from or recombination to the valence level or trapping centres. In the case of collision ionization due to another conduction electron, thermal excitation due to the absorption of phonons may play a part. In the case of recombination, the energy lost by the recombining electron may be emitted as phonons or photons, or alternatively given to another conduction electron.

(5) Diffusion due to spatial variation of parameters describing the electron distribution function.

The resulting Boltzmann equation is lengthy even when written in a formal manner (cf. Heller 1951, Franz 1956, and Stratton 1961), and so complex as to be quite intractable if the terms describing the various processes are presented explicitly. The techniques used in practical application depend on ignoring some terms in the Boltzmann equation and approximating to others. Since we are considering bulk effects, the diffusion term under (5) is ignored. A further simplifying circumstance arises from the fact that collision ionization and its inverse affect only those electrons of very high and very low energy, at least in so far as these processes are concerned with the valence band. There is a large range of energy in which the electron has too little energy to ionize, and yet sufficient to make a recombining collision relatively improbable. For this energy range, only terms under (1), (2), and (3) will remain in the Boltzmann equation. These considerations have led to two main attacks on the problem.

The first method, which is valid for low densities of conduction electrons, commences by ignoring the electron–electron collisions. In the intermediate range of energies, the distribution function is then determined entirely by the applied field and the electron–phonon collisions. This calculation was done by Fröhlich (1947b) who showed that it is impossible to normalize the distribution so obtained unless those electron–electron collisions are included that cause ionization and recombination. The inclusion of these processes has been attempted by Heller (1951), Franz (1956, 1952), and Veelken (1955).

The second method, which is valid for high densities of conduction electrons, begins by assuming that the electron–electron collisions are of such great importance that they determine the form of the distribution function and in fact cause it to be Maxwellian (cf. Fröhlich 1947a, 1952, O'Dwyer 1954, 1957, Fröhlich and Paranjape 1956, Stratton 1957, 1958, and Paranjape 1961). The influences of the applied field and the electron–phonon collisions are then required to determine the parameters of the distribution, viz. its temperature (which is different from the lattice temperature), and the shift of the distribution in momentum space.

A more general procedure based on direct solution of the Boltzmann equation was given by Yamashita and Watanabe (1954) for the case of scattering by acoustic modes only; it is pointed out by Conwell (1967) that practical application of their result would be limited to very pure crystals at low temperatures.

We shall first discuss the low-density and high-density approximations for polar crystals, for which the interaction of conduction electrons is primarily with longitudinal optical lattice modes. For non-polar crystals both acoustical and optical modes may contribute to electron scattering, and the results are complex; they will be discussed briefly.

(a) The low-density approximation for polar crystals

The energy distribution for the conduction electrons in the low-density approximation was considered by Fröhlich (1947b) and Franz (1952). In accordance with the previous discussion, we consider only the energy range for which

$$\hbar\omega \ll E < I, \tag{4.12}$$

where ω is a lattice vibrational frequency, and I the energy required to cause collision ionization from the valence band. Using a Legendre expansion for the electron distribution function (θ is the angle between the direction of the field and the electron momentum)

$$f(E,\ \theta,\ F) = f_0(E,\ F) + f_1(E,\ F)\cos\theta..., \tag{4.13}$$

Fröhlich (1947b) showed that, for high field strengths, the symmetrical part of the distribution function is not Maxwellian. Using the continuum model for electron–phonon interaction and (2.57) for the relaxation time, Stratton (1958) gives the result

$$f_0(E,\ F) \propto \exp\left[\left\{\frac{1-\exp(\hbar\omega/k_0 T)}{\hbar\omega}\right\}\times\right.$$

$$\left.\times \int^E \frac{dE}{1+\dfrac{2F^2E^2}{3\ln(4E/\hbar\omega)}\left(\dfrac{\epsilon^*\{\exp(\hbar\omega/k_0 T)-1\}}{m^*e\omega}\right)^2}\right]. \tag{4.14}$$

For non-zero values of the field strength, $f_0(E,\ F)$ does not tend to zero as E tends to infinity, so that the distribution function cannot be normalized for any applied field. It should be emphasized that this result depends on the assumption that the only mechanism preventing electrons from reaching high energies is the electron–phonon collision. When the electron energy exceeds I, collision ionization becomes a very probable process and its effects would have to be invoked to make possible the normalization of the density function.

However, one may use eqn (4.14) as a hopeful approximation for low fields and for values of the electron energy not too much greater than a lattice quantum. This has been done by Stratton

(1958), who has shown that (4.14) can be simplified in a certain special case. To this end we define a dimensionless field strength

$$F_0 = m^* e \omega / \epsilon^* \hbar \qquad (4.15)$$

characteristic of a given polar crystal (it is of order 0·3 MV cm^{-1} for a typical alkali halide). Then for values of the electron energy such that $E \gg \hbar\omega$, and field strengths such that $F < F_0$, eqn (4.14) becomes

$$f_0(E, F) \propto \left\{ 1 + \frac{2}{9 \ln (4k_0 T / \hbar\omega)} \left(\frac{F}{F_0} \frac{E}{k_0 T}\right)^2 \right\} \exp(-E/k_0 T).$$

$$(4.16)$$

In the limiting case of zero field strength, (4.16) is Maxwellian, and the use of eqns (2.55) and (2.57) in eqn (2.70) yields

$$\langle \mu(0) \rangle = \frac{4\sqrt{2}}{3\pi^{\frac{1}{2}}} \frac{\hbar^2 \epsilon^*}{em^{*\frac{3}{2}}} \frac{1}{(k_0 T)^{\frac{1}{2}}}, \qquad (4.17)$$

which is identical with the mobility derived from eqn (2.72). If the non-Maxwellian distribution (4.16) is used for finite values of F, the mean value of the relaxation time must then be calculated from the more general result (2.69). The corresponding field-dependent mobility is found to be

$$\langle \mu(F) \rangle = \langle \mu(0) \rangle \{ 1 + F^2 / 2F_0^2 \ln (4k_0 T / \hbar\omega) \}. \qquad (4.18)$$

It will be shown below that this is of similar form to the mobility variation to be expected in the high-density case; it appears that electron–electron collisions do not greatly affect the mean mobility for the conditions under which eqn (4.16) was derived.

(b) *The high-density approximation for polar crystals*

The high-density approximation for polar crystals was considered by Fröhlich and Paranjape (1956), and is based on the assumption that conduction electrons are so numerous that they interchange energy with each other much more rapidly than with the lattice. It is assumed that this dominance of electron–electron collisions causes the electron distribution function to be Maxwellian, with a temperature T_e which may be different from the lattice temperature T; the effect of the field is simply to shift the electron distribution function in momentum space. The distribution function can therefore be written

$$f(\mathbf{p}) \propto \exp(-|\mathbf{p}-\mathbf{p}_0|^2/2m^*k_0T_e), \qquad (4.19)$$

where $\mathbf{p}_0$ is proportional to $\mathbf{F}$ and subject to the inequality

$$p_0^2/m^* \ll k_0T_e. \qquad (4.20)$$

The Boltzmann equation for the distribution function $f(\mathbf{p})$ in momentum space for steady-state conditions is

$$\left(\frac{\partial f}{\partial t}\right)_{\mathrm{F}} + \left(\frac{\partial f}{\partial t}\right)_{\mathrm{L}} + \left(\frac{\partial f}{\partial t}\right)_{\mathrm{e}} = 0, \qquad (4.21)$$

where the subscripts F, L, and e refer to rates of change due to applied field, lattice, and other electrons respectively. Electron–electron collisions conserve both energy and momentum so that

$$\left.\begin{aligned} \sum_{\mathbf{p}} \mathbf{p}\left(\frac{\partial f}{\partial t}\right)_{\mathrm{e}} &= 0 \\ \sum_{\mathbf{p}} \frac{p^2}{2m^*}\left(\frac{\partial f}{\partial t}\right)_{\mathrm{e}} &= 0 \end{aligned}\right\} \qquad (4.22)$$

From (4.21) and (4.22) it follows that

$$\sum_{\mathbf{p}} \mathbf{p}\left\{\left(\frac{\partial f}{\partial t}\right)_{\mathrm{F}} + \left(\frac{\partial f}{\partial t}\right)_{\mathrm{L}}\right\} = 0 \qquad (4.23)$$

and

$$\sum_{\mathbf{p}} \frac{p^2}{2m^*}\left\{\left(\frac{\partial f}{\partial t}\right)_{\mathrm{F}} + \left(\frac{\partial f}{\partial t}\right)_{\mathrm{L}}\right\} = 0, \qquad (4.24)$$

from which equations the unknown parameters $\mathbf{p}_0$ and T_e of the distribution function (4.19) can be determined.

Taking first the terms due to the applied field, which is assumed to be in the z direction, we have

$$\left(\frac{\partial f}{\partial t}\right)_{\mathrm{F}} = \frac{\partial f}{\partial p_z}\frac{\mathrm{d}p_z}{\mathrm{d}t} = -eF\frac{\partial f}{\partial p_z}, \qquad (4.25)$$

there being no changes in the x and y directions on account of the field. Changing the summations into integrations we have

$$\int \mathbf{p}\left(\frac{\partial f}{\partial t}\right)_{\mathrm{F}} \mathrm{d}\mathbf{p} = -eF\int p_z\left(\frac{\partial f}{\partial p_z}\right)\mathrm{d}\mathbf{p} = -eFn \qquad (4.26)$$

where n is the total electron density given by

$$n = \int f(\mathbf{p})\,\mathrm{d}\mathbf{p}. \tag{4.27}$$

Also

$$\int \frac{p^2}{2m^*}\left(\frac{\partial f}{\partial t}\right)_{\mathrm{F}} \mathrm{d}\mathbf{p} = -\frac{eF}{2m^*}\int p^2 \frac{\partial f}{\partial p_z}\,\mathrm{d}\mathbf{p}$$

$$= -\frac{eF}{m^*}\int p_z f\,\mathrm{d}\mathbf{p} = -\frac{eF}{m^*}p_0 n. \tag{4.28}$$

If we then turn to the terms due to interaction with the lattice, a consideration of all collision processes gives the formal result

$$\left(\frac{\partial f}{\partial t}\right)_{\mathrm{L}} = -\sum_{\mathbf{q}} \{f(\mathbf{p})P^{\mathrm{a}}(\mathbf{p},\,\mathbf{p}+\mathbf{q})-f(\mathbf{p}+\mathbf{q})P^{\mathrm{e}}(\mathbf{p}+\mathbf{q},\,\mathbf{p})+$$

$$+f(\mathbf{p})P^{\mathrm{e}}(\mathbf{p},\,\mathbf{p}-\mathbf{q})-f(\mathbf{p}+\mathbf{q})P^{\mathrm{a}}(\mathbf{p}-\mathbf{q},\,\mathbf{p})\}, \tag{4.29}$$

in which $\mathbf{q}$ is related to the phonon wave number $\mathbf{w}$ by $\mathbf{q} = \hbar\mathbf{w}$. The emission and absorption probabilities are given by eqns (2.35), (2.37), and (2.38) with interaction constants for a polar crystal expressed by (2.45) or (2.46). If $f(\mathbf{p})$ is developed in a series of Legendre functions with the field direction as axis,

$$f(\mathbf{p}) = f_0(\mathbf{p})+f_1(\mathbf{p})\cos\theta+..., \tag{4.30}$$

then using eqn (4.19) and the inequality (4.20) we have

$$f_0 \propto \exp(-p^2/2m^*k_0 T_{\mathrm{e}}) \tag{4.31}$$

and

$$f_1 = -p_0 \frac{\partial f_0}{\partial p} = \frac{pp_0}{m^*k_0 T}f_0. \tag{4.32}$$

Using eqns (4.30), (4.31), and (4.32) we can then write

$$\left(\frac{\partial f}{\partial t}\right)_{\mathrm{L}} = g_0(p)+p_0\cos\theta\,g_1(p)+... \tag{4.33}$$

and determine g_0 and g_1 from the integral form of (4.29) with the distribution functions expanded to first order in p_0. The functions g_0 and g_1 have been given by Fröhlich and Paranjape (1956) in a manner which involves the interaction constant $G(\mathbf{w})$ and the phonon spectrum $\omega(\mathbf{w})$ in an arbitrary way; the temperature T_{e} is also involved from the expansion of the distribution function (4.19) in powers of p_0, and the temperature

T from the expression for the number of phonons in eqn (2.37). From eqn (4.33) we then have

$$\int p_z \left(\frac{\partial f}{\partial t}\right)_{\mathrm{L}} \mathrm{d}\mathbf{p} = \frac{p_0}{3} \int p g_1(p) \, \mathrm{d}\mathbf{p} \qquad (4.34)$$

and

$$\int \frac{p^2}{2m^*} \left(\frac{\partial f}{\partial t}\right)_{\mathrm{L}} \mathrm{d}\mathbf{p} = \int \frac{p^2}{2m^*} g_0(p) \, \mathrm{d}\mathbf{p}, \qquad (4.35)$$

and substituting eqns (4.26), (4.28), (4.34), and (4.35) into eqns (4.23) and (4.24) we have

$$eFn = \tfrac{1}{3} p_0 \int p g_1(p) \, \mathrm{d}\mathbf{p} \qquad (4.36)$$

and

$$eFnp_0 = \tfrac{1}{2} \int p^2 g_0(p) \, \mathrm{d}\mathbf{p}. \qquad (4.37)$$

Elimination of p_0 from eqns (4.36) and (4.37) yields

$$e^2 F^2 n^2 = \tfrac{1}{6} \int p^2 g_0(p) \, \mathrm{d}\mathbf{p} \int p g_1(p) \, \mathrm{d}\mathbf{p} \qquad (4.38)$$

which is the basic relation expressing the electron temperature T_e in terms of the applied field F. Elimination of F from eqns (4.36) and (4.37) then gives

$$p_0^2 = \frac{3}{2} \frac{\int p^2 g_0(p) \, \mathrm{d}\mathbf{p}}{\int p g_1(p) \, \mathrm{d}\mathbf{p}}, \qquad (4.39)$$

which yields p_0 provided that T_e is already determined from (4.38).

The mean mobility can now be found from

$$\langle \mu \rangle = p_0/m^* F. \qquad (4.40)$$

In this, it is convenient to follow the work of Stratton (1958), who found analytic expressions for various special cases. Making the definitions

$$\left. \begin{aligned} x &= \hbar\omega/k_0 T \\ x_e &= \hbar\omega/k_0 T_e \end{aligned} \right\} \qquad (4.41)$$

we find for the limiting value of the mobility at low field strengths

$$\langle \mu(0) \rangle = \frac{3\sqrt{(2\pi)}}{4} \frac{\hbar \epsilon^*}{e\omega} \left(\frac{\hbar\omega}{m^{*3}}\right)^{\frac{1}{2}} \frac{1}{n_{\mathrm{w}}} \{x^{\frac{3}{2}} \exp(x/2) K_1(x/2)\}^{-1}, \qquad (4.42)$$

where K_1 is a modified Bessel function. This result is valid for all temperatures; in the high-temperature limit it is equal to (4.17) except for a numerical factor of order unity.

For non-zero field, Stratton calculated the mobility using a Taylor expansion and obtained

$$\langle\mu(F)\rangle = \langle\mu(0)\rangle\{1+A(x)F^2/F_0^2\}, \qquad (4.43)$$

where F_0 is a field strength characteristic of the material given by (4.15), and the temperature-dependent quantity $A(x)$ is given by

$$A(x) = \frac{3\pi}{8}\frac{\{K_1(x/2)-2xK_0(x/2)\}}{n_w^2 x^3\,\exp(x)K_0(x/2)K_1^2(x/2)}. \qquad (4.44)$$

This expression can be simplified in the high-temperature and low-temperature limits to yield

$$A(x) \simeq 3\pi/16\ln(D/x) \quad \text{for} \quad x \ll 1 \qquad (4.45)$$

and

$$A(x) \simeq -(\tfrac{3}{4}x)\exp(2x) \quad \text{for} \quad x \gg 1, \qquad (4.46)$$

where D is a number defined in terms of Euler's constant, γ, by

$$\ln D = \ln 4 - \gamma \simeq 0.809.$$

The quantity $A(x)$ is shown as a function of $1/x$ in Fig. 4.5. It will be seen that mobility variations are by far the greatest

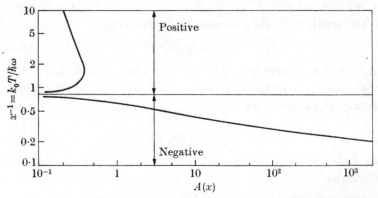

FIG. 4.5. Plot of the temperature dependence of $A(x)$ in the mobility equation (4.43) (after Stratton 1958).

at low temperatures, and that $A(x)$ changes sign for

$$k_0 T/\hbar\omega \simeq 0\cdot82.$$

The negative sign of $A(x)$ is a reflection of the physical fact that very low-energy electrons encounter more friction with the ionic lattice as they are warmed up; on the other hand, high-energy electrons are less impeded by the lattice as their energy increases.

Equations (4.43) and (4.45) for the high-temperature approximation to the field-dependent mobility in the region of high electron densities can be compared with eqn (4.18) for the mobility at low electron densities. The expressions are similar and lead to the conclusion that electron–electron collisions do not markedly affect mobility variations at high temperatures.

The condition for the validity of the Taylor expansion (4.43) is $F \ll F_0/|A|^{\frac{1}{2}}$; since for a typical alkali halide $F_0 \sim 0\cdot3$ MV cm^{-1}, the expansion should be good up to very high field strengths except at low temperatures (cf. Fig. 4.5).

It is of interest to investigate the conduction-electron density that would be required for the above considerations to be valid. Consider electrons with thermal energies at some temperature greater than the lattice temperature, but of the same order. The ratio of the probabilities of emission and absorption of a lattice quantum by an electron is $(1+n_w)/n_w$, provided the electron energy exceeds $\hbar\omega$, so that the net energy loss is $1/(1+2n_w)$ times the energy transferred per collision. If $\tau(E, T)$ is the average time between electron–phonon collisions for electrons of energy E and a lattice at temperature T, then the rate of energy transfer from such an electron to the lattice is given approximately by

$$\left(\frac{\mathrm{d}E}{\mathrm{d}t}\right)_{\mathrm{L}} \simeq \frac{\hbar\omega}{\tau(E,\,T)}\frac{1}{1+2n_w}. \tag{4.47}$$

On the other hand, the rate of energy loss by such an electron due to collisions with other electrons is given approximately by

$$\left(\frac{\mathrm{d}E}{\mathrm{d}t}\right)_{\mathrm{e}} \simeq \frac{4\pi n e^{*4}}{(2m^*E)^{\frac{1}{2}}} \tag{4.48}$$

(cf. Pines 1953), where n is the conduction-electron density, and e^* is an effective electronic charge differing from e by a factor depending on the dielectric constant. Thus, in so far as energy transfer is concerned, eqns (4.47) and (4.48) show that electron-electron and electron–phonon collisions are of equal importance at a critical density

$$n_c \simeq \frac{(2m^*E)^{\frac{1}{2}}}{4\pi e^{*4}} \frac{\hbar\omega}{\tau(E, T)} \frac{1}{1+2n_w}. \tag{4.49}$$

For higher densities, electron–electron collisions should predominate, and the distribution function will be given by eqn (4.19). The order of magnitude of the critical electron density can be found by assuming $(1+2n_w) \sim 1$ and using eqns (2.55) and (2.57) in (4.49) to give

$$n_c \sim \frac{m^*e^2\omega^2}{4\pi e^{*4}} \frac{1}{\epsilon^*}. \tag{4.50}$$

Substitution of typical values shows that $n_c \sim 10^{17}$ cm^{-3} for the alkali halides. Stratton (1958) has also examined the case in which the lattice temperature is low (i.e. $k_0 T \ll \hbar\omega$), a circumstance which at first sight appears to invalidate the above argument, since it would be expected that most electrons could not then emit a phonon. However, it is shown that this is not so for high values of the field, and that (4.50) gives the correct critical density under these conditions, since the electron temperature is such that $k_0 T_e \gtrsim \hbar\omega$ even if the lattice temperature T is very low.

The condition that the number of conduction electrons be greater than the critical density estimated in (4.50) ensures only the Maxwellian form of the distribution function; the relation (4.32) for the asymmetrical part of the distribution function requires the more stringent condition that the rate of exchange of momentum amongst electrons be large in comparison with that between electrons and lattice vibrations. Stratton (1958) has shown that this normally requires higher electron densities than those estimated above. However, the form of (4.32) is preserved, irrespective of electron density (as in (2.67) for the low-density case), so that it should remain a good first approximation even if critical densities are exceeded.

(c) Non-polar crystals

The case of non-polar crystals is more complex for various reasons. In the first place, both optical and acoustical modes may contribute appreciably to electron scattering in non-polar crystals. Secondly, the question of critical electron densities is more difficult; and lastly various ranges of applied field strength must be distinguished (cf. Stratton 1957).

The question of critical density can be dealt with in an approximate way, by seeking a condition for the dominance of electron–electron collisions in determining the distribution function without distinguishing between its symmetrical and asymmetrical components. Assuming that the energy exchanged per collision with the phonons is of the same order as the maximum required by conservation of momentum, we can say that an electron of wave number k exchanges energy of order $\hbar s k$ per collision, where s is the velocity of sound. Writing this in terms of the electron energy and using eqns (2.61), (4.47), and (4.48), we obtain for the critical electron density

$$n_{\mathrm{c}} \simeq \frac{1}{50}\frac{C^2 m^{*3}E^2}{\hbar^4 e^{*4}MN}. \tag{4.51}$$

Insertion of typical values gives $n_{\mathrm{c}} \sim 10^{14}\ \mathrm{cm^{-3}}$, which is a density several orders of magnitude lower than the corresponding quantity for polar crystals.

The basic equations giving the conduction-electron temperature and the drift of the Maxwellian distribution in the direction of the field are again (4.38) and (4.39). The various possible formulae for the mobility have been considered at length by Stratton (1957) and tabulated in detail by him. We shall however consider only a relatively simple case first treated by Shockley (1951), which includes the interaction with acoustic modes alone, and assumes that the electron temperature is not too much greater than the lattice temperature. The resulting mobility as a function of field strength is

$$\langle \mu(F) \rangle = \langle \mu(0) \rangle \sqrt{2} [1 + \{1 + (3\pi/8)(\langle \mu(0) \rangle F/s)^2\}^{\frac{1}{2}}]^{-\frac{1}{2}} \quad (4.52)$$

(cf. Conwell 1967). For low field strengths $\langle \mu(0) \rangle$ is given by eqn (2.63) and (2.71) with a $T^{-\frac{3}{2}}$ temperature dependence; for high fields, (4.52) approximates to

$$\langle \mu(F) \rangle \simeq 1 \cdot 3 (s \langle \mu(0) \rangle / F)^{\frac{1}{2}}, \quad (4.53)$$

which has a $T^{-\frac{3}{4}}$ temperature dependence.

The mobility form (4.53) was verified for solid argon by Miller, Howe, and Spear (1968), whose results are shown in Fig. 4.6. The $F^{-\frac{1}{2}}$ dependence of mobility was observed over a wide range of field strength in solid argon. Results by the same authors on solid krypton and xenon showed only small intervals in which the $F^{-\frac{1}{2}}$ mobility dependence was followed; in all

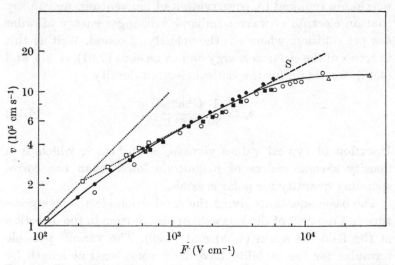

FIG. 4.6. The field-strength dependence of the electron drift velocity in solid A at 82 K. The dotted line is the extrapolated linear region, and the dashed curve (marked S) is calculated from (4.53). The differently marked points correspond to different inter-electrode spacings. (After Miller et al. 1968.)

cases the drift velocity saturated to a constant value at very high field strengths so that in this region the mobility shows an F^{-1} dependence. This saturation of the drift velocity has also been observed by Pruett and Broida (1967), who suggest that it is due to inelastic scattering by impurities.

(d) Field-dependent polaron mobility

Very little has been written on field-dependent polaron mobility. For the large Fröhlich polaron discussed in § 6(a) the result for the low-field mobility is essentially similar to that for a free electron in an ionic crystal, and one would expect explicit field dependence to follow the lines of the discussion in § 4.2(a).

There is no theoretical work on the field-dependent mobility of the hopping polaron, but Emtage (1971) has given a calculation of the effect of a strong field in inducing a transition from the band to the hopping mode of transport within the framework of the Holstein model discussed in § 2.3(b). The band for small polarons at low temperatures may be extremely narrow, so that even a moderate electric field produces a potential difference between adjacent atoms that is of order of the polaron band width. Under these circumstances, the electron wave function is localized, and conduction more accurately described as hopping. Emtage (1971) has computed the current density against field strength characteristic of the Holstein model for a particular set of parameters, and his results are shown in Fig. 4.7. The sequence of events anticipated as field strength is increased is first, a low-field ohmic region of small-polaron band conduction, then an intermediate-field transition region of negative resistance or constant current, and finally a high-field ohmic region of hopping conduction. The transition region appears most pronounced for long mean free paths, since in this case the low-field conduction is most accurately described as band conduction. No experimental results support this theory, since there is not even unambiguous evidence of small-polaron band conduction.

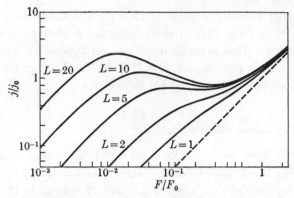

Fig. 4.7. Current density against field strength in dimensionless units for a Holstein model polaron. The parameter L is the mean free path in units of interatomic distance. In the notation of § 2.3(b),

$$\gamma = 7$$
$$J = \hbar\omega = 300k_0$$

were assumed values for constant parameters. For the various mean free paths listed on the diagram, temperatures, field scale factors, and current scale factors are given by the following table.

L	1	2	5	10	20
$k_0 T/\hbar\omega$	0·38	0·35	0·31	0·29	0·26
F_0 (V cm⁻¹)	1600	2000	2500	2900	3300
j_0 (arbitrary units)	1·0	0·77	0·51	0·35	0·25

(After Emtage 1971.)

4.3. Enhancement of electronic carrier density at high fields

Enhanced electronic conductivity at high fields may occur because of multiplication of charge carriers, and this effect may be alternative or additional to the change in mobility discussed in the preceding section. Electronic-charge multiplication in the bulk must be due to ionization either from traps or from the valence band; the latter is usually regarded as a prelude to electrical breakdown, and we will therefore confine our attention in this section to ionization from trapping levels. There exist two order-of-magnitude theories which correspond essentially to high-density and low-density approximations. These will be treated in turn.

(a) The Fröhlich amorphous-solid model

A model for the electronic structure of a dielectric solid introduced by Fröhlich (1947a) has been illustrated in Fig. 2.19, and the appropriate low-field conductivity has been discussed in § 2.4(a). In order to find the explicit field dependence of the conductivity, one needs to calculate the electron temperature T_e for the high-field case, in which it can no longer be set equal to the lattice temperature T. To calculate this temperature in the steady state requires that the energy gained by the conduction electrons from the applied field be equated to the energy that the electrons transfer to the lattice.

The energy gain from the applied field is due only to the free C electrons and is

$$A(F, T_e, T) = \sigma F^2 = n_D e \mu \gamma F^2 \exp(-W/k_0 T_e), \quad (4.54)$$

where eqn (2.111) has been used for the conductivity. Compared with the exponential term, it is assumed that the mobility μ is a slowly varying function of temperature.

The energy loss to the lattice is caused either by inelastic collision of a C electron with a phonon, or by phonon emission coupled with electron transitions between various isolated levels. Transitions to D levels are very improbable (since they require a multiphonon process) compared with transitions between the relatively closely spaced S levels. The inequality (2.109) then implies that the process mainly responsible for transferring energy from the field to the lattice is phonon emission associated with electron transitions between S levels. Note that the field transfers energy direct to the C electrons, which share it with the S electrons in maintaining a common temperature T_e; the latter are then mainly responsible for the energy transfer to the lattice. The energy transfer is

$$B(T_e, T) = \hbar\omega \sum (W_e - W_a), \quad (4.55)$$

where W_e is the number of transitions per second of an electron to some energy E involving phonon emission, and W_a is the corresponding quantity from energy E involving absorption. The summation in (4.55) is over all S-level energies. If $P(E)$ is the electron transition probability (which is not a function of temperature), we may write

$$W_a = P(E)f_0(E)n(T) \tag{4.56}$$

and

$$W_e = P(E)f_0(E+\hbar\omega)\{n(T)+1\}, \tag{4.57}$$

where f_0 is the Maxwellian distribution function, and $n(T)$ is the number of lattice quanta at the temperature T. Let $1/\tau_S$ be the average value of $P(E)$ in the sense

$$n_S/\tau = \sum P(E)f_0(E) \tag{4.58}$$

where the summation is over S-level energies. Then using the relation

$$n_S = n_D \exp(-W/k_0T_e + \Delta V/k_0T_e) \tag{4.59}$$

for the density of electrons in shallow isolated levels together with eqns (2.110), (4.56), (4.57), and (4.58) in (4.55), we obtain

$$B(T_e, T) = \frac{\hbar\omega n_C}{\tau_S\gamma} n(T)\exp(\Delta V/k_0T_e) \times$$
$$\times \{\exp(\hbar\omega/k_0T - \hbar\omega/k_0T_e) - 1\}. \tag{4.60}$$

Equating (4.54) with (4.60) as the condition for a steady state we obtain

$$DF^2 \exp(-\Delta V/k_0T_e) = \exp(\hbar\omega/k_0T - \hbar\omega/k_0T_e) - 1, \tag{4.61}$$

where D is given by

$$D = e\mu\tau_S\gamma/\hbar\omega n(T), \tag{4.62}$$

which can be considered a relatively slowly varying function of temperature. The temperature of the electron distribution can then be determined from eqn (4.61), and this can be done by algebraic approximation in certain cases. The principle of the solution is illustrated in Fig. 4.8. For sufficiently low field strengths such as F_1 there are two solutions for the temperature, T_1 and T_2, the former of which is readily seen to be stable and

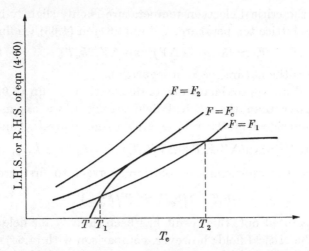

Fig. 4.8. Diagram to illustrate the solution of eqn (4.60). The family of curves plotted for various values of F represents the L.H.S. and the single curve the R.H.S. of this equation. Intersections give possible solutions for the electron temperature.

the latter unstable; we therefore identify T_1 with the electron distribution temperature T_e. When the field increases to a critical value F_c only one temperature T_c is possible, while for fields $F_2 > F_c$ the electron temperature should increase without limit.

The field dependence of the conductivity is most easily expressed in terms of this critical field strength F_c. In order to obtain it algebraically from eqn (4.61) we note from Fig. 4.8 that $F = F_c$ if the derivatives of both sides of (4.61) with respect to T_e are equal, with $T_e = T_c$, i.e.

$$\Delta V D F_c \exp(-\Delta V/k_0 T_c) = \hbar\omega \exp(\hbar\omega/k_0 T - \hbar\omega/k_0 T_c). \quad (4.63)$$

Combining eqn (4.63) with eqn (4.61) (the latter with $F = F_c$ and $T = T_c$), we find

$$(1 - \hbar\omega/\Delta V)\exp(\hbar\omega/k_0 T - \hbar\omega/k_0 T_c) = 1,$$

and in view of eqn (2.108) this becomes

$$1/k_0 T - 1/k_0 T_c \simeq 1/\Delta V, \quad (4.64)$$

so that the critical electron temperature is only slightly higher than the lattice temperature. Using (4.64) in (4.63) we find

$$F_c \simeq (\hbar\omega/e \; D \; \Delta V)^{\frac{1}{2}} \exp(\Delta V/2k_0 T) \qquad (4.65)$$

where e is the natural base of logarithms.

To find an approximate algebraic solution of eqn (4.61) for fields lower than the critical field strength, we expand the right-hand side of (4.61) and use (4.65) to eliminate D. This gives

$$(1/\epsilon\Delta V)F^2/F_c^2 \exp(\Delta V/k_0 T - \Delta V/k_0 T_e) \simeq 1/k_0 T - 1/k_0 T_e, \quad (4.66)$$

to which the approximate solution (correct to first order in F^2/F_c^2) is

$$1/k_0 T - 1/k_0 T_e \simeq F^2/F_c^2 \, \Delta V. \qquad (4.67)$$

As we pointed out, this result has been derived for fields well below the critical field; however, comparison with (4.64) shows that it passes over into the same result as that valid at the critical field strength. We therefore adopt (4.67) as being approximately correct for all values of the applied field below the critical.

Use of eqn (2.111) for the conductivity then yields

$$\sigma(F) = \sigma(0)\exp\left(\frac{W}{\Delta V}\frac{F^2}{F_c^2}\right). \qquad (4.68)$$

No direct evidence has been given for this square-law variation with field strength for the logarithm of the conductivity; the result (4.68) has however been used in calculations of current–voltage characteristics, as will be shown in § 5.4.

(b) The Poole–Frenkel effect

The name 'Poole–Frenkel effect' has been applied to describe a current–voltage dependence that is linear on a Schottky plot, but is supposed to arise from field-dependent thermionic emission from traps in the bulk of the insulator. The effect of the field in lowering the thermionic work function of a trapped electron for a one-dimensional model is shown in Fig. 4.9. If the trapping potential is a fixed Coulomb potential, then the potential energy to the right of the trap in an applied field F will be given by

$$eV(x) = -eFx - e^2/\epsilon x. \qquad (4.69)$$

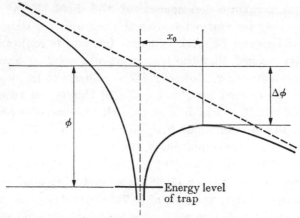

F𝗂𝗀. 4.9. One-dimensional potential-energy diagram illustrating the Poole–Frenkel effect. The work function in the absence of a field is ϕ, and $\Delta\phi$ is the reduction in work function caused by the applied field.

The maximum value of the potential energy occurs at a point distant x_0 to the right of the trap, where

$$x_0 = (e/F\epsilon)^{\frac{1}{2}} \tag{4.70}$$

and the work function is reduced by an amount

$$\Delta\phi = 2(e^3F/\epsilon)^{\frac{1}{2}}. \tag{4.71}$$

This result differs from that of eqn (3.4) by a factor 2, since in this case the attracting charge is in a fixed position.

Frenkel (1938) then estimated the electrical conductivity in the presence of a strong field as follows. In the absence of an electric field the number of free electrons present in the conduction band because of thermal ionization from traps will be proportional to $\exp(-\phi/2k_0T)$ if the Fermi level is midway between the trapping levels and the bottom of the conduction band. Hence the electrical conductivity in the presence of a field should be proportional to

$$\exp\{-(\phi-\Delta\phi)/2k_0T\},$$

or

$$\sigma(F) = \sigma(0)\exp(\Delta\phi/2k_0T)$$
$$= \sigma(0)\exp\{(e^3F/\epsilon)^{\frac{1}{2}}/k_0T\}. \tag{4.72}$$

The temperature dependence of the Poole–Frenkel conductivity may be contrasted with the temperature dependence of the Richardson–Schottky current (cf. eqn (3.52)), and it is frequently alleged that the temperature-dependent exponents differ by a factor of 2. However, if one adheres to the argument as originally quoted by Frenkel (1938) this is not true, since the $\Delta\phi$ of eqn (3.4) which is used in the Richardson–Schottky equation (3.52) is one-half that given by eqn (4.71) and used in the Poole–Frenkel equation (4.72). This point has been raised by Seki (1970), Simmons (1967), Mark and Hartman (1968), and Yeargan and Taylor (1968); the factor appearing in the denominator of the exponent of (4.72) depends on the position of the Fermi level. For an n-type material with no acceptor sites, conditions will be as envisaged in Frenkel's original argument, and eqn (4.72) is appropriate. In the presence of compensation, deep-lying acceptor levels will have the effect of lowering the Fermi level, and altering the statistics involved in the derivation of eqn (4.72). For the limit in which the concentration of conduction electrons is very much less than both the donor and the acceptor densities, eqn (4.72) will be replaced by

$$\sigma(F) = \sigma(0)\exp(\Delta\phi/k_0 T)$$
$$= \sigma(0)\exp\{2(e^3 G/\epsilon)^{\frac{1}{2}}/k_0 T\}. \qquad (4.73)$$

In this case the exponent of the temperature-dependent term does differ by a factor 2 from that in the Richardson–Schottky equation (3.52). The magnitude of the exponent in question depends upon the degree of compensation present in the insulator, and may vary between the limiting cases expressed by eqns (4.72) and (4.73).

In addition to the uncertainty concerning the statistical mechanics of the model, the validity of the model itself is open to serious question. Equation (4.71) has been derived on the basis of a Coulomb trap in a dielectric continuum having a single dielectric constant ϵ; this is such an oversimplified model that the final result must be considered as an order-of-magnitude calculation—indeed it seems very clear from Frenkel's original paper that he intended no more than this. Jonscher

(1967) has, however, reviewed the subject in some depth and claims that the usual explanation is substantially correct, but that it should not be interpreted too literally. Elsewhere, Jonscher and Ansari (1971) point out that the proper distinction to draw between the Poole–Frenkel and Richardson–Schottky mechanisms is whether the effect is bulk or electrode-dominated; there is no compelling reason to judge this from the magnitude of a factor in the exponent of the temperature-dependent term.

There is in principle an additional point of distinction between the two effects. Equation (3.52) for Schottky emission predicts a straight line on a graph $\log j$ against $F^{\frac{1}{2}}$, while eqn (4.72) predicts a straight line on a graph $\log \sigma$ against $F^{\frac{1}{2}}$. Under most circumstances, experimental results can be plotted as a good straight line on either type of graph; the mechanisms cannot therefore be separated by this means.

The measurements of the current–voltage characteristics of silicon nitride films by Sze (1967) are shown in Fig. 4.10 and are typical of experimental results that have been ascribed to Poole–Frenkel conduction. Sze analyzed his results on the basis of eqn (4.73), which applies to a highly compensated substance; use of eqn (4.71) then yields a value of 5·5 for the dielectric constant, which is of a reasonable order of magnitude.

An empirical modification of the Poole–Frenkel law was proposed by Hartman, Blair, and Bauer (1966), whose results for silicon monoxide are shown in Fig. 4.11. The equation proposed for the current density as a function of voltage was

$$j = A_1 d \exp(A_2 V^{\frac{1}{2}} - \phi/k_0 T)\{1 - \exp(-A_2^2 V)\}, \quad (4.74)$$

where

$$A_2 = d^{-\frac{1}{2}}\{(\psi/k_0 T) + \eta\}. \quad (4.75)$$

The quantities A_1, ψ, and η are certain constants, d is the film thickness, and ϕ is the activation energy for low-field conductivity. From a plot of the zero-field intercepts on an Arrhenius diagram, it was found that $\phi = 0.430$ eV. The slope of the lines in Fig. 4.11 yields a value $\epsilon = 3$ for the dielectric constant if one assumes uncompensated material, rising to $\epsilon = 12$ if a high degree of compensation is present.

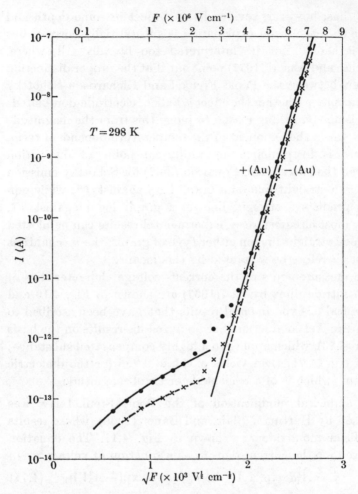

FIG. 4.10. Current against square root of field for a silicon nitride film at room temperature. The electrodes were gold and molybdenum with polarity as indicated on the diagram. The film thickness was 2900 Å and the area $1·6 \times 10^{-4}$ cm². (After Sze 1967.)

Silicon monoxide has also been investigated by Servini and Jonscher (1969), who worked over a much wider range of temperature and applied voltage than previous authors. Their results for one of their more highly conducting samples are shown in Fig. 4.12. Four different conduction regions are

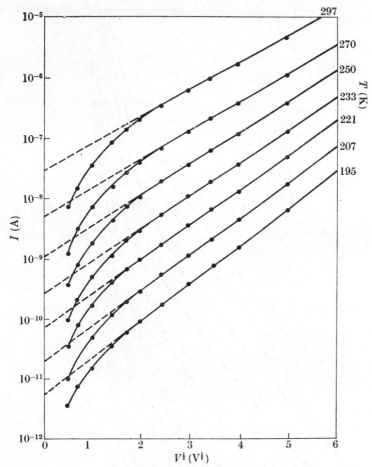

Fig. 4.11. Current–voltage characteristics of a 7080 Å thick SiO film for various temperatures, with aluminium electrodes. The full lines are fitted to eqn (4.74). (After Hartman *et al.* 1966.)

distinguished and separated approximately by the dotted lines on the diagram. In region I, conduction is ohmic, and is considered to be ionic. In region II the results give straight lines on a Schottky diagram as shown in Fig. 4.13. A fit of these data to eqn (4.72) for uncompensated material yields for the dielectric constant $\epsilon = 3$. Servini and Jonscher (1969) were uncertain as to the value to be assigned to the activation energy,

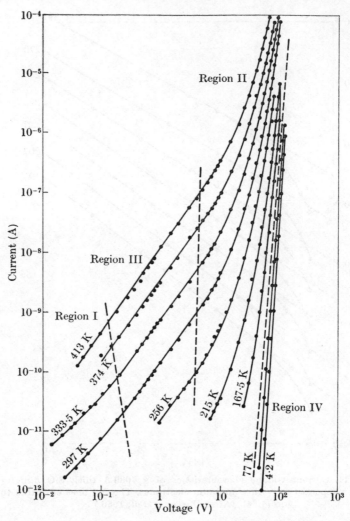

FIG. 4.12. Current–voltage characteristics of a 3800 Å thick silicon monoxide film for various temperatures, with aluminium electrodes. The significance of the various regions is explained in the text. (After Servini and Johscher 1969.)

since their data produced a curved line on an Arrhenius plot. For low fields, their experimental data appeared to converge towards a value $\phi \simeq 0.3$ eV, while extrapolation techniques gave a value approximately twice as large. Region III of Fig. 4.12 is represented by a power law $I \propto V^{1.5}$, and region IV is

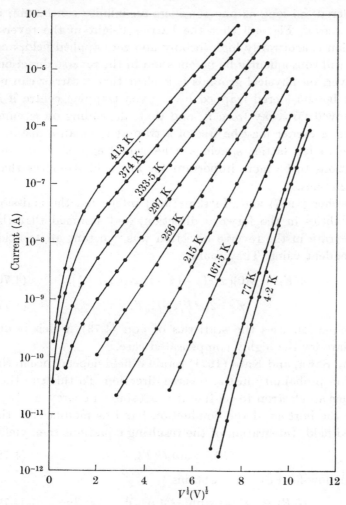

FIG. 4.13. Schottky plot of data from region II of Fig. 4.12 (after Servini and Jonscher 1969).

virtually temperature independent and characterized by very steep power law current–voltage characteristics. These results will be discussed more fully below.

Various refinements have been proposed to the Poole–Frenkel effect. One set of amendments endeavours to overcome the restrictions implied in the one-dimensional nature of the model, and in the fact that only forwards emission is usually considered. So long as one considers an infinite continuum, it is clear from Fig. 4.9 that the barrier height in the reverse direction is arbitrarily high for any non-zero applied field, and there will consequently be no emission in the reverse direction. However, on physical grounds it is clear that a carrier can no longer be considered trapped by a given trapping centre if it has moved far away from it, and that, depending on circumstances, a carrier may be free of a trap at as small a distance as five to ten lattice spacings. One would expect these considerations to be more important for low field strengths than for high ones.

Jonscher (1967) and Hartke (1968) integrated the emission probabilities in the forward direction, and assumed that the release rate in the reverse direction was constant at its field-independent value. They found

$$\sigma(F) = \sigma(0)[\alpha^{-2}\{1+(\alpha-1)\exp(\alpha)\}+\tfrac{1}{2}], \qquad (4.76)$$

where

$$\alpha = 2(e^3F/\epsilon)^{\frac{1}{2}}/k_0T. \qquad (4.77)$$

This equation uses the statistics of eqn (4.73), which is appropriate for the highly compensated case.

Idea, Sawa, and Kato (1971) retain a field dependence in the emission probability in the reverse direction. To this end they consider an electron to be free if it attains an energy $\delta \sim k_0T$ below the bottom of the conduction band as modified by the applied field. Integration of the resulting equations then yields

$$\sigma(F) = \sigma(0)G(F). \qquad (4.78)$$

For the low-field case one obtains

$$G(F) = (4\gamma/\alpha^2)\sinh(\alpha^2/4\gamma) \quad \text{if} \quad \alpha \leq 2\gamma \qquad (4.79)$$

and

$$\gamma = \delta/2k_0T \qquad (4.80)$$

where α is defined by eqn (4.77). For the high-field case,

$$G(F) = \alpha^{-2}\{(\alpha-1)\exp(\alpha-\gamma)-2\gamma\exp(-\alpha^2/4\gamma)+\exp(\gamma)\}$$
$$\text{if} \quad \alpha \geq 2\gamma. \qquad (4.81)$$

Ieda *et al.* (1971) fitted these equations to the experimental data of Hartman *et al.* shown in Fig. 4.11; they obtained a fit equally as good as that found using the empirical formula (4.74).

Equations (4.79) and (4.81) use the statistics of the original Frenkel calculation appropriate to the uncompensated case. A plot of the dimensionless conductivity as a function of field strength is shown in Fig. 4.14 as determined from eqns (4.78),

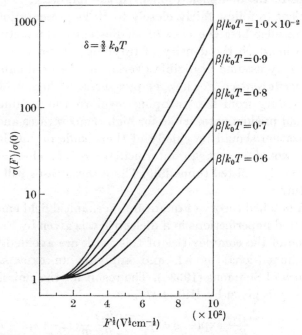

FIG. 4.14. Conductivity–field-strength characteristics determined from eqns (4.78), (4.79), and (4.81). The parameter $\beta/k_0 T$ of eqn (4.82) is given in units of cm$^{\frac{1}{2}}$ V$^{-\frac{1}{2}}$. (After Idea et al. 1971.)

(4.79), and (4.81). The curves are shown for various values of the parameter

$$\lambda = \beta/k_0 T = 2e^{\frac{3}{2}}/\epsilon^{\frac{1}{2}}k_0 T. \qquad (4.82)$$

This means that, apart from a small effect due to variation of the square root of the dielectric constant, the parameter λ of the curves in Fig. 4.14 is inversely proportional to absolute temperature.

By far the most complete reassessment of the Poole–Frenkel

effect has been presented by Hill (1971), who considers single-centre and multi-centre conduction, tunnelling, and thermally assisted tunnelling in addition to the other modifications already discussed. Hill then applies his analysis to the experimental results of various workers, and shows that it is possible to interpret all the measurements as being various limiting cases of the one basic model. In the following, it will be convenient to follow fairly closely to Hill's presentation.

Consider first the case of single-centre conduction, i.e. the case in which the density of trapping centres is so small that one may assume negligible overlap of the Coulomb field from adjacent centres. For low temperatures and high field strengths, tunnelling from the trapping level to the conduction band should predominate, while for high temperatures and low fields ionization should be chiefly of thermionic origin; in addition, one would expect an intermediate region characterized by thermally assisted tunnelling. These three cases will be treated in turn.

A detailed theory of quantum-mechanical field emission from isolated imperfections in a dielectric was given by Franz (1956). Some of the complexities of this work are avoided by using a one-dimensional model, and applying an expansion of the theory of Simmons (1963a). The result for the emission current density is given by Hill (1971) as

$$j = \frac{N_i v_i e}{B^2} \exp(-A\bar{\phi}^{\frac{1}{2}}) \frac{\pi B k_0 T}{\sin(\pi B k_0 T)} \{2 - \exp(-B\phi)\},$$

$$(4.83)$$

in which the product $N_i v_i$ is treated as an assignable parameter; this reflects the difficulty of calculating a supply function appropriate to the emission of electrons from traps. The remaining quantities in (4.83) are defined as follows,

$$A\bar{\phi}^{\frac{1}{2}} = \frac{\sqrt{2}(2m^*)^{\frac{1}{2}}\phi^{\frac{3}{2}}}{\hbar e F} \lambda \left\{1 - \frac{\gamma F}{2\lambda} \ln\left(\frac{1+\lambda}{1-\lambda}\right)\right\}^{\frac{1}{2}}, \quad (4.84)$$

$$B = \frac{2(m^*\phi)^{\frac{1}{2}}}{\hbar e F} \left\{1 - \frac{\gamma F}{2\lambda} \ln\left(\frac{1+\lambda}{1-\lambda}\right)\right\}^{-\frac{1}{2}}, \quad (4.85)$$

$$\gamma = \beta^2/\phi^2, \quad (4.86)$$

and
$$\lambda = (1-\gamma F)^{\frac{1}{2}} = \{1-(\Delta\phi/\phi)^2\}^{\frac{1}{2}}, \tag{4.87}$$

from eqns (4.61) and (4.82). Hill's presentation as expressed in these equations clearly brings out the analogy with the theory of field emission from a plane metallic electrode described in §3.1. The temperature dependence of the current follows at once from eqn (4.83) as

$$j(T) = j(0)\,\frac{\pi Bk_0 T}{\sin(\pi Bk_0 T)}\,, \tag{4.88}$$

which for low temperatures, $\pi Bk_0 T \lesssim 1$, becomes, on expansion of the donominator,

$$\frac{j(T)-j(0)}{j(0)} = \frac{(\pi Bk_0 T)^2}{6}\,. \tag{4.89}$$

Turning now to the case of purely thermionic emission, we face again the difficulty of making the transition from a calculation that is valid in the equilibrium situation to one that we hope describes the steady state. In Frenkel's original calculation (eqn (4.72)), this was achieved by finding a factor describing the ratio of the steady-state conductivity $\sigma(F)$ to its value $\sigma(0)$ in the limit of the approach to equilbrium. However, Hill (1971) attempts an absolute calculation of the current density in the steady state, and this requires a knowledge of the velocity of a carrier in the field direction when it has attained sufficient energy to clear the trap. Two different assumptions are made; if the host material is crystalline it is assumed that the velocity in the field direction is independent of energy and given by

$$v_x = \mu_1 F, \tag{4.90}$$

while for an amorphous material it is assumed that

$$v_x = \mu_2 F^{\frac{1}{2}}. \tag{4.91}$$

It should be noted that μ_1 is a mobility while μ_2 is a quantity having no simple physical interpretation—it is certainly not a mobility. Two approximations have already been described to deal with emission in the reverse direction; Hill introduces another by assuming that the barrier in the reverse direction will be increased by the same amount as the barrier in the forward direction has been decreased, i.e. by $\beta F^{\frac{1}{2}}$, where β is the Poole–Frenkel constant given by eqn (4.82). It is argued that this assumption gives zero current at zero field, and is to this extent a reasonable one. The one-dimensional calculation therefore needs to be modified in that

$$2 \sinh \alpha \quad \text{replaces} \quad \exp(\alpha), \tag{4.92}$$

while in a three-dimensional calculation

$$2\alpha^{-2}(\alpha \cosh \alpha - \sinh \alpha) \quad \text{replaces} \quad \exp(\alpha), \tag{4.93}$$

where α is given by (4.77).

The emission current density is now found from

$$j = \int_{\phi-\beta F^{\frac{1}{2}}}^{\infty} N(E)ev_x \, dE, \tag{4.94}$$

where $N(E)$ is the density of carriers per unit energy range and is written (for highly compensated material)

$$N(E) = N_i k_0 T \exp(-E/k_0 T). \tag{4.95}$$

The use of (4.90) or (4.91), (4.92) or (4.93), and (4.95) in (4.94) then yields four possible formulae for the current density. For a crystalline host material we have

$$j = 2eN_i(k_0 T)^4 \beta^{-2} \mu_1 \exp(\phi/k_0 T)\alpha^2 \sinh \alpha \tag{4.96}$$

for one-dimensional emission, while for the three-dimensional case

$$j = 2eN_i(k_0 T)^4 \beta^{-2} \mu_1 \exp(-\phi/k_0 T)(\alpha \cosh \alpha - \sinh \alpha). \tag{4.97}$$

For an amorphous material, the one-dimensional calculation yields

$$j = 2eN_i(k_0 T)^3 \beta^{-1} \mu_2 \exp(-\phi/k_0 T)\alpha \sinh \alpha, \tag{4.98}$$

while the three-dimensional result is

$$j = 2eN_i(k_0 T)^3 \beta^{-1} \mu_2 \exp(-\phi/k_0 T)\alpha^{-1}(\alpha \cosh \alpha - \sinh \alpha). \tag{4.99}$$

These four equations can be reduced to a two-variable system

$$\mathscr{J} = f(\alpha, \sinh \alpha), \tag{4.100}$$

where

$$\mathscr{J} = jT^{-n} \exp(\phi/k_0 T) \tag{4.101}$$

with $n = 3$ or 4 depending on whether one makes assumption (4.91) or (4.90) respectively. The nature of the function f in eqn (4.100) depends on the choice made in both sets of assumptions and is given explicitly in eqns (4.96) to (4.99).

These equations describing Poole–Frenkel emission under various circumstances have been fitted to experimental data in an interesting manner, as will be shown below, but they appear to be open to a serious conceptual objection. The operation indicated by eqns (4.92) and (4.93) implies that carriers are emitted in various directions with a probability that depends only on the energy barrier they face, while the substitution of (4.90) or (4.91) into (4.94) implies emission only in the forward direction. It therefore seems that formulae (4.96)–(4.99) have been derived on inconsistent assumptions.

The case of thermally assisted tunnelling is treated approximately by consideration of a sequential process consisting of thermal excitation to some energy ϕ_0 followed by tunnelling into conduction levels. Combining the exponential factors for tunnelling and thermal excitation, one expects

$$j \propto \exp(-A\bar{\phi}_0^{\frac{1}{2}} - \phi_0/k_0 T), \qquad (4.102)$$

where $A\bar{\phi}_0^{\frac{1}{2}}$ is defined on the analogy of (4.84) and can be simplified to the approximate form

$$A\bar{\phi}_0^{\frac{1}{2}} = \frac{\sqrt{2}(2m^*)^{\frac{1}{2}}}{\hbar e F} \{(\phi - \phi_0) - \beta F^{\frac{1}{2}}\}(\phi - \beta F^{\frac{1}{2}})/\phi^{\frac{1}{2}} \quad (4.103)$$

The thermal excitation energy that corresponds to maximum current flow is given by

$$\frac{\partial}{\partial \phi_0}(-A\bar{\phi}_0^{\frac{1}{2}} - \phi_0/k_0 T) = 0.$$

Solution gives

$$\phi_{0max} = \left\{\frac{\sqrt{2}(2m^*)^{\frac{1}{2}}k_0 T}{\hbar e F}\right\}^{\frac{2}{3}} (\phi - \beta F^{\frac{1}{2}})^{\frac{4}{3}} \qquad (4.104)$$

for $\phi_0 \lesssim \phi - \beta F^{\frac{1}{2}}$; and, if one assumes that the bulk of the current flow corresponds to thermal excitation to this energy followed by tunnelling, the result is

$$j \propto \exp\left\{ - \frac{2\phi_{0max}}{k_0 T} \left(1 - \frac{\phi_{0max}}{2(\phi - \beta F^{\frac{1}{2}})} \right) \right\}. \qquad (4.105)$$

For low fields one expects that ϕ_{0max} will approach $\phi - \beta F^{\frac{1}{2}}$, i.e. the emission will be chiefly thermionic, and (4.105) simplifies with the use of (4.104) to

$$j \propto \exp\left[\left\{ \frac{\sqrt{2(2m^*)^{\frac{1}{2}}}}{\hbar e F} \right\}^{\frac{2}{3}} (k_0 T)^{-\frac{1}{3}} \right]. \qquad (4.106)$$

Hill (1971) applies the above analysis to the experimental data of various investigators. The results of Servini and Jonscher (1969), which are shown in Fig. 4.12, can be replotted to test agreement with Hill's theory. The data shown in Fig. 4.15 are for temperatures greater than 140 K and correspond to region II of Fig. 4.12 with a lot of overlap into regions I and III at the lower end of the curve. The two curves plotted in Fig. 4.15 are eqns (4.96) and (4.97), which are appropriate to crystalline host material. Clearly in this case there is little to choose between a one-dimensional and a three-dimensional theory. The values of parameters deduced from the fit of data are

$$\left. \begin{array}{l} \phi = 0 \cdot 35 \text{ eV} \\ \beta = 1 \cdot 32 \times 10^{-4} \text{ eV V}^{-\frac{1}{2}} \text{ cm}^{\frac{1}{2}} \\ N_i \mu_1 = 10^{16} \text{ cm}^{-1} \text{ V}^{-1} \text{ s}^{-1} \text{ eV}^{-2} \end{array} \right\}. \qquad (4.107)$$

With the parameters ϕ and β from (4.107), region IV of Fig. 4.12 was analysed in terms of the tunnelling theory of eqns (4.83) and (4.89). The results are shown in Fig. 4.16 and show excellent agreement between experiment and theory. The disposable parameter in this fit of data is

$$N_i v_i = 3 \times 10^9 \text{ cm}^{-2} \text{ s}^{-1} \text{ eV}^{-2}. \qquad (4.108)$$

Finally the data in the intermediate region (region III of Fig. 4.12) are checked against eqn (4.106) as shown in Fig. 4.17. The straight-line portion of the characteristics corresponds to thermally assisted tunnelling; the changing gradient at low temperatures marks the approach to the region of temperature-independent tunnelling.

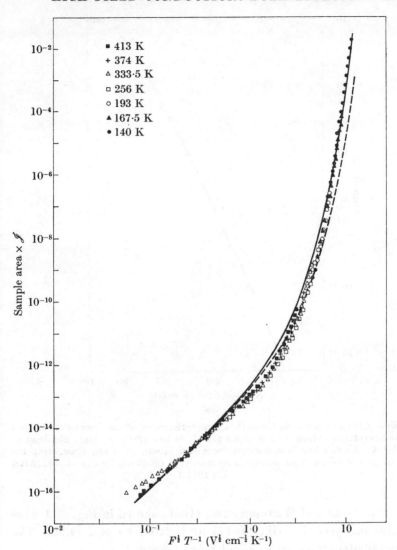

FIG. 4.15. Current–field-strength characteristics of silicon monoxide for high temperatures obtained by Servini and Jonscher (1969). The ordinate is proportional to $\mathscr{J} = jT^{-4}\exp(0\cdot35/k_0T)$ with k_0T in eV, and the abscissa $F^{\frac{1}{2}}T^{-1}$ is proportional to α of eqn (4.77). The full line is eqn (4.96) and the dashed line (4.97). (After Hill 1971.)

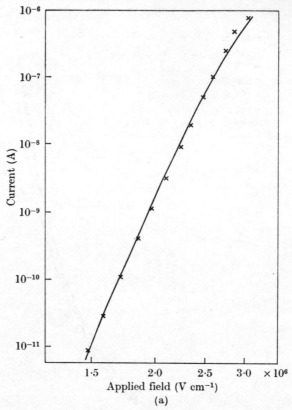

FIG. 4.16. Current–field-strength characteristics of silicon monoxide for low temperatures obtained by Servini and Jonscher 1969. (a) Data obtained at 4·2 K; the full line is calculated from eqn (4.83). (b) Data illustrating the square-power-law dependence on temperature predicted by eqn (4.89). (After Hill 1971.)

The results of Hartman *et al.* (1966) shown in Fig. 4.11 were also analysed by Hill (1971) and fitted to eqn (4.99). The parameters required to effect this fit were

$$\left.\begin{aligned}
\phi &= 0\cdot36 \text{ eV} \\
\beta &= 1\cdot56 \times 10^{-4} \text{ eV V}^{-\frac{1}{2}} \text{ cm}^{\frac{1}{2}} \\
N_i\mu_1 &= 6\cdot6 \times 10^{18} \text{ cm}^{-1} \text{ V}^{-1} \text{ s}^{-1} \text{ eV}^{-2}
\end{aligned}\right\}, \qquad (4.109)$$

which compare favourably with those given in (4.107). Hill's

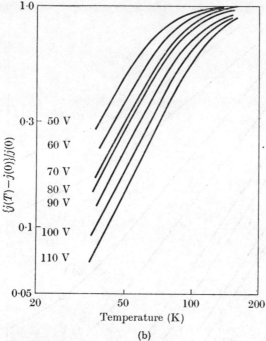

FIG. 4.16. (*continued*)

analysis of these experimental results from different sources leads to almost the same value of activation energy in both cases; the evaluation of ϕ from an Arrhenius plot leads to uncertainties and different values. However, the Poole–Frenkel constant β is substantially the same whether it is evaluated by Hill's method or straight from a Schottky plot of the results.

The case of multiple-centre conduction was also treated by Hill (1971). Emission of a charge carrier from a centre occurs into an adjacent centre rather than into the conduction band. For a one-dimensional model, Hill found

$$j = 2eN_i s(k_0 T)^2 \nu \exp(-\phi'/k_0 T)\sinh \alpha', \qquad (4.110)$$

where

$$\alpha' = eFs/2k_0 T \qquad (4.111)$$

and s is the distance between adjacent centres. The derivation of (4.110) is based on the same considerations as those leading

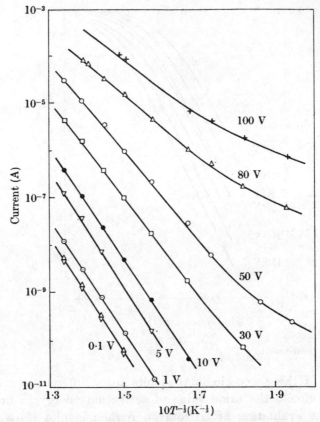

FIG. 4.17. Current–temperature characteristics of silicon monoxide for various field strengths obtained by Servini and Jonscher 1969. The lines on the plot are in accord with the application of eqn (4.106) to region III of Fig. 4.12.

to (4.1) for field-dependent ionic mobility, and it too can be cast in the form of (4.100) and (4.101), with α' substituted for α and $n = 2$. The results of Klein and Lisak (1966) for silicon monoxide are shown in Fig. 4.18 together with the interpretation in terms of eqn (4.110). The parameters required to fit the data were

$$\left.\begin{array}{l} \phi' = 0\cdot35 \text{ eV} \\ s = 20 \text{ Å} \end{array}\right\}. \tag{4.112}$$

The activation energy for transitions between adjacent traps

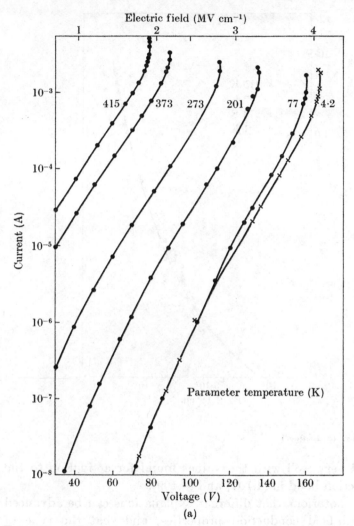

FIG. 4.18. (a) Current–voltage characteristics of silicon monoxide for various temperatures (after Klein and Lisak 1966). (b) The same data plotted with ordinate proportional to $\mathcal{J} = jT^{-2}\exp(0\cdot35/k_0T)$ with k_0T in eV, and abscissa proportional to α' of eqn (4.111). The full line is eqn (4.110). (After Hill 1971.)

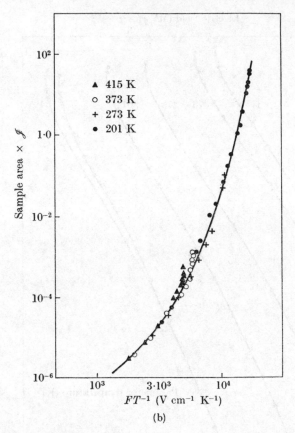

FIG. 4.18. (*continued*)

agreed very well with the values found for activation to the conduction band in (4.107) and (4.109).

It is notorious that different explanations can be advanced for high-field conduction properties, and that the type of explanation depends in large measure on the plots chosen to exhibit the experimental results. One must therefore accept all interpretations with a certain amount of reserve.

HIGH-FIELD CONDUCTION: COMBINED ELECTRODE AND BULK EFFECTS

5.1. Introduction

VARIOUS models of current flow through dielectrics between electrodes have been proposed, and the complicating factors that one may consider produce quite insoluble problems. However, much progress has been made for certain special models, and we list below the simplifying assumptions which will be adhered to in the remainder of this chapter.

(1) Plane parallel electrodes will be assumed, so that the problem and its solution will be formulated in an essentially one-dimensional manner.

(2) Steady-state processes only will be considered. The total current density will therefore be independent of both position and time.

(4) Poisson's equation will be used to relate charge density to the spatial variation of the field. This ignores the particulate nature of the elementary charges and may not always be justifiable at very low current densities.

(5) Empirical relations will be used to describe either collision ionization or electron–hole recombination where appropriate.

(6) Charge-carrier diffusion due to concentration gradient will be neglected.

It is not necessary to make all of these assumptions to make progress in any given case. Thus Lampert and Mark (1970) give an extensive discussion of transients for the case of ohmic contacts; the effect of various geometries is also discussed. However the restrictions are useful when discussing other types of electrode contacts.

Within the framework of the above assumptions, the basic equations pertaining to the dielectric can be set up. The current density will be given by

$$j = j_n + j_p = ne\mu_n F + pe\mu_p F = \text{const.}, \tag{5.1}$$

where j_n and j_p are the electron and hole components of the current density, and n and p the electron and hole carrier densities. Poisson's equation is

$$\partial F/\partial x = 4\pi e(p-n)/\epsilon, \tag{5.2}$$

where x is the distance coordinate measured in the direction of the field, and ϵ is the dielectric constant. The continuity equation may be written

$$\partial j_n/\partial x = -\partial j_p/\partial x, \tag{5.3}$$

and the space variation may be ascribed to processes such as recombination or collision ionization.

It will be shown below that suitable manipulation of eqns (5.1)–(5.3) leads to a non-linear differential equation for the field strength; the current-injecting properties of the electrodes play the role of boundary conditions to this differential equation. It is possible to set the problem up in dimensionless form with parameters characteristic of the dielectric if the boundary conditions are homogeneous. There are two possible types of homogeneous boundary conditions, viz:

(1) Ohmic contacts are specified mathematically by

$$F_{\text{cath}} = 0 \tag{5.4a}$$

regardless of the electron emission current, and

$$F_{\text{an}} = 0 \tag{5.4b}$$

regardless of the hole emission current. These conditions are of course limiting situations applying approximately to contacts that emit charge carriers copiously.

(2) Blocking contacts are specified mathematically by

$$j_{n\text{cath}} = 0 \tag{5.5a}$$

regardless of the field at the cathode, and

$$j_{p\text{an}} = 0 \tag{5.5b}$$

regardless of the field at the anode. Although these conditions are on the components of current density, they nevertheless serve as homogeneous boundary conditions to the differential equation for the field strength, since they effectively exclude physical parameters of the electrodes.

The boundary conditions are non-homogeneous if the emission current from either electrode is a function of the field strength, i.e.

$$j_{n\text{cath}} = j_n(F_{\text{cath}}) \tag{5.6a}$$

for electron emission from the cathode, and

$$j_{p\text{an}} = j_p(F_{\text{an}}) \tag{5.6b}$$

for the hole emission from the anode. The subscripts 'an' and 'cath' refer in all cases to the values of quantities adjacent to the anode and cathode respectively.

5.2. Ohmic contacts

The study of the current–voltage characteristics of dielectrics with ohmic electrode contacts was originally made by Mott and Gurney (1948) and Rose (1955), and developed and reviewed by Lampert (1964) and Lampert and Mark (1970). Since the very extensive research in this field is described in detail in the last reference, we shall simply consider some of the more basic results.

(a) Insulator with single-species injection

We assume the injected species to be electrons, noting that exactly the same results hold for injected holes. We shall consider in turn a trap-free insulator and an insulator with traps for the case in which thermal ionization from traps is not dependent on field strength; finally we consider field-dependent ionization from traps.

If there are neither thermally generated carriers nor traps in the insulator, the model is the solid-state analogue of the thermionic vacuum diode. The difference is that the electric field in the insulator is assumed to move the injected charge with a velocity determined by its mobility, whereas in a vacuum the field moves the charge with a constant acceleration. The governing equations become, from (5.1) and (5.2),

$$j = ne\mu_n F \tag{5.7}$$

and

$$\partial F/\partial x = -4\pi ne/\epsilon. \tag{5.8}$$

Elimination of the number of density of carriers from (5.7) and (5.8) yields

$$F(\partial F/\partial x) = -4\pi j/\epsilon\mu_n. \tag{5.9}$$

Integration of (5.9) with the boundary condition (5.4a) and a cathode–anode spacing of L leads to

$$j = (9/32\pi)\epsilon\mu_n(V^2/L^3), \tag{5.10}$$

which is usually referred to as the Mott and Gurney square law, or Child's law for solids. (Child's law for a vacuum diode predicts $j \propto V^{\frac{3}{2}}$ rather than $j \propto V^2$.)

Let us now drop the restriction on thermally generated carriers and traps in the insulator, and consider the somewhat more realistic situation in which both are present. We first present a simplified theory due to Lampert (1956), which outlines the general features of the solution to the problem by examining certain limiting cases. Consider an insulator containing traps and thermally excited electrons. Let n_c be the density of thermally excited electrons in the bulk neutral insulator with no applied voltage, and let N_t be the density of traps that are uncharged in the unoccupied state. The general character of the current–voltage characteristics is deduced with reference to the Lampert triangle shown in Fig. 5.1.

The lower curve corresponds to Ohm's law for the neutral crystal,

$$j = n_c e\mu_n F, \tag{5.11}$$

where F is taken to be the mean electric field V/L. The actual current–voltage characteristic cannot lie below this curve, since the injection of electrons at the ohmic contact can only add additional carriers to those already present thermally.

The upper curve is Child's law given by eqn (5.10); it is valid for a trap-free crystal with a negligible density of thermally generated carriers. The actual current–voltage curve

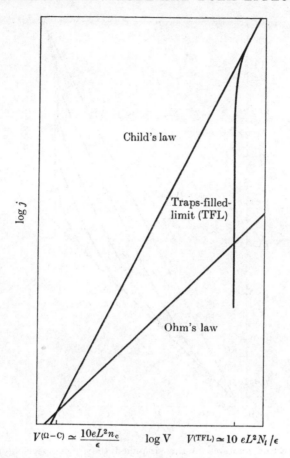

Fig. 5.1. The Lampert triangle showing limiting current–voltage relation-ships for space-charge-limited currents in an insulator with traps (after Lampert 1956).

cannot lie above this line (unless the mobility increases with increasing field strength), since it represents the case in which all of the injected excess charge is in the conduction band; if any of the injected charge were trapped the current would be lower. The cross-over point for which ohmic and Child's law currents are equal occurs for the voltage

$$V^{(\Omega-C)} = 32\pi eL^2 n_c/9\epsilon \simeq 10eL^2 n_c/\epsilon, \tag{5.12}$$

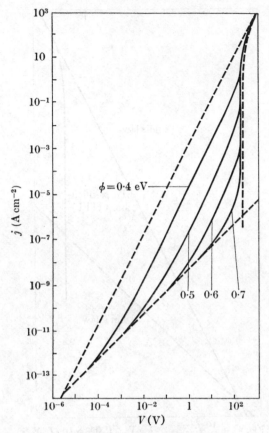

FIG. 5.2. Exact solutions to the simplified theory of space-charge-limited current flow in CdS with a single discrete trapping level. The dashed curves are the boundaries of the Lampert triangle, and the full curves are exact solutions corresponding to trap depths as indicated. The data used were

$$\epsilon = 11,$$
$$\mu_n = 200 \text{ cm}^2 \text{ V}^1 \text{ s}^{-1},$$
$$T = 300 \text{ K},$$
$$L = 5 \times 10^{-3} \text{ cm},$$
$$N_t = 10^{14} \text{ cm}^{-3},$$

and the Fermi level was taken to be $0 \cdot 75$ eV below the conduction band, leading to $n_c = 10^6 \text{ cm}^{-3}$. (After Lampert 1956.)

which is given by elimination of j from eqns (5.10) and (5.11).

The vertical curve to the right of the diagram in Fig. 5.1 corresponds to the potential drop across the insulator if all traps are filled with electrons. Using Poisson's equation, one finds readily

$$V^{(\text{TFL})} = 2\pi e L^2 N_t/\epsilon \simeq 10 e L^2 N_t/\epsilon. \tag{5.13}$$

The actual current–voltage curve cannot lie to the right of the traps-filled-limit curve and below the Child's law curve, since the traps-filled-limit curve represents the most unfavourable possible distribution of charge in so far as conduction is concerned.

It is assumed in this analysis, as the approximate values of the two limiting voltages in Fig. 5.1 make clear, that

$$N_t/n_c \gg 1.$$

This condition holds true in many insulators over a wide range of temperatures. The actual current–voltage characteristic must then lie within the Lampert triangle shown in Fig. 5.1; this is illustrated by some exact solutions to the simplified theory for CdS crystals given by Lampert (1956) and shown in Fig. 5.2. It is clear from the figure that, provided the traps are well above the Fermi level, the actual current–voltage characteristic possesses a large square-law region in which the current is less than the Child's law current by a constant factor. This factor is the ratio of the free charge to the total charge, and is given by

$$\theta_0 = (N_c/N_t)\exp(-\phi/k_0 T), \tag{5.14}$$

where ϕ is the depth of the single-trap level, and N_c is the effective density of states in the conduction band. For the region within the Lampert triangle the square-law characteristic becomes, from eqns (5.10) and (5.14),

$$j - \frac{9}{32\pi} \epsilon \mu_n \frac{V^2}{L^3} \frac{N_c}{N_t} \exp\left(-\frac{\phi}{k_0 T}\right). \tag{5.15}$$

The general current–voltage characteristics predicted by Lampert's theory have been observed in many insulating substances to which ohmic contact has been made; details are given by Lampert and Mark (1970).

It has been pointed out by Murgatroyd (1970a) that there is a deviation from the square-law characteristic (5.15) if the probability of ionization from traps is field-dependent. Murgatroyd assumed a Poole–Frenkel type factor in the ionization probability, so that the ratio of free charge to total charge becomes

$$\theta = \theta_0 \exp(\beta F^{\frac{1}{2}}/k_0 T) \tag{5.16}$$

in place of (5.14); β is the Poole–Frenkel constant defined by (4.82). Eliminating n_c and θ from eqns (5.8), (5.11), and (5.16), we get

$$j = \frac{\theta_0 \epsilon \mu_n}{4\pi} F \exp\left(\frac{\beta F^{\frac{1}{2}}}{k_0 T}\right) \frac{\partial F}{\partial x}. \tag{5.17}$$

Although this equation can be integrated to yield distance as a function of field strength, the functional relation cannot be analytically inverted and integrated again to give the current–voltage characteristic. However, Myrgatroyd developed a dimensionless computation which led to an approximate analytical result

$$j = \frac{9}{32\pi} \epsilon \mu_n \frac{V^2}{L^3} \frac{N_c}{N_t} \exp[-\{\phi - 0 \cdot 891 \beta (V/L)^{\frac{1}{2}}\}/k_0 T]. \tag{5.18}$$

Note that for a given mean field strength eqn (5.18) predicts a current density inversely proportional to the thickness. This feature would provide good confirmatory evidence of the postulated mechanism, but does not appear to have been observed in those cases for which thickness dependence was measured (cf. Hartman et al. 1966).

(b) Insulator with double injection

If electrons are injected from the cathode and holes from the anode, the space charge from each species tends to cancel, so that a large current density does not necessarily imply a large net charge density; this in turn does not imply a large rate of variation of field strength, and large current densities may be produced with low voltages. The limiting mechanism is recombination of electrons and holes either directly from conduction to valence levels, or indirectly via recombination centres. Various circumstances of the problem are discussed by Lampert and Mark (1970), who also outline an exact solution for the case of a perfect trap-free insulator with no recombination centres which was given by Parmenter and Ruppel (1959). Since this solution parallels another for the case of blocking contacts we give a brief discussion of it.

In addition to the current equation (5.1) and Poisson's equation (5.2), the relation (5.3) is now given specifically by

$$\partial j_n/\partial x = -\partial j_p/\partial x = e\langle v\sigma_R\rangle np, \qquad (5.19)$$

where v is the magnitude of the relative electron–hole velocity, and σ_R is the velocity-dependent recombination cross-section. The brackets $\langle\ \rangle$ represent an average over the velocity distribution of electrons and holes. The solution is expressed in terms of suitably defined quantities; the recombination coefficient is expressed as a mobility

$$\mu_R = \epsilon\langle v\sigma_R\rangle/8\pi e, \qquad (5.20)$$

and the electron and hole mobilities put in dimensionless form by the definitions

$$\left.\begin{array}{l} \nu_n = \mu_n/\mu_R \\ \nu_p = \mu_p/\mu_R \end{array}\right\}. \qquad (5.21)$$

For ohmic contacts (boundary conditions (5.4)), Parmenter and Ruppel (1959) find

$$j = (9/32\pi)\epsilon\mu_{\text{eff}}(V^2/L^3), \qquad (5.22)$$

which is of the same form as (5.10) except that the electron mobility is replaced by an effective mobility given by

$$\mu_{\text{eff}} = \mu_R\nu_n\nu_p\left[\frac{2}{3}\frac{\{\frac{3}{2}(\nu_n+\nu_p)-1\}!}{(\frac{3}{2}\nu_n-1)!(\frac{3}{2}\nu_p-1)!}\right]^2 \times$$
$$\times \left[\frac{(\nu_n-1)!(\nu_p-1)!}{(\nu_n+\nu_p-1)!}\right]. \qquad (5.23)$$

Two particularly interesting cases arise for the limits of very large and very small recombination coefficients. A very large recombination coefficient is implied by the conditions

$$v_n \ll 1 \quad \text{and} \quad v_p \ll 1.$$

Equation (5.23) then reduces to

$$\mu_{\text{eff}} = \mu_p + \mu_n, \tag{5.24}$$

which corresponds to pure electron space-charge-limited current from the cathode and pure hole space-charge-limited current from the anode, with the two currents meeting and annihilating within the insulator. A very small recombination coefficient is implied by the conditions

$$v_n \gg 1 \quad \text{and} \quad v_p \gg 1.$$

This case corresponds to the injected plasma with $n \simeq p$ almost everywhere. If Stirling's approximation is used for the factorials, eqn (5.23) reduces to

$$\mu_{\text{eff}} = \tfrac{2}{3}[2\pi\{\mu_n\mu_p(\mu_n+\mu_p)/\mu_R\}]^{\frac{1}{2}}. \tag{5.25}$$

The potential functions appropriate to various cases are shown in Fig. 5.3. In all cases the field is zero at the electrodes and rises to a maximum at some point within the insulator.

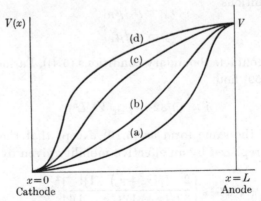

FIG. 5.3. Potential distribution for double injection in a perfect insulator with ohmic contacts at anode and cathode. (a) $\mu_p \ll \mu_R$ and μ_n. (b) $\mu_n > \mu_p \gg \mu_R$. (c) $\mu_p > \mu_n \gg \mu_R$. (d) $\mu_n \ll \mu_R$ and μ_p. (After Parmenter and Ruppel 1959.)

The whole subject of space-charge-limited current flow is treated in detail by Lampert and Mark (1970) and will not be further pursued here.

5.3. Blocking contacts

Blocking contacts are specified by the boundary conditions (5.5), which, in effect, set the total injected current at zero. It is possible, however, that there may be a non-zero steady-state current if there is collision ionization within the insulator; electron current is then ejected at the anode and hole current at the cathode. This problem of double ejection was investigated by O'Dwyer (1968). The equations to be solved are the current equation (5.1), Poisson's equation (5.2), and the collision ionization equation. We shall assume that the electrons only cause collision ionization, and that the holes produced are of relatively very low mobility and not causing collision ionizations. In addition, recombination will be neglected compared with collision ionization. Equation (5.3) can then be written

$$\partial j_n/\partial x = -\partial j_p/\partial x = -e\alpha n, \qquad (5.26)$$

where

$$\alpha = \alpha_n \exp(-H/F), \qquad (5.27)$$

and α_n is a constant ionization coefficient with the dimension of an inverse time, and H is a constant field strength. (For a discussion of this formula see §7.2(b).)

To eliminate the carrier densities and work only in terms of field strength, current densities, and their derivatives, we solve eqns (5.2) and (5.26) for n and p, yielding

$$n = -(\alpha e)^{-1}\partial j_n/\partial x \qquad (5.28)$$

and

$$p - (\epsilon/4\pi e)\partial F/\partial x - (\alpha e)^{-1}\partial j_n/x. \qquad (5.29)$$

Substitution of eqn (5.29) in eqn (5.1) then yields

$$j = j_n + \frac{\epsilon\mu_p}{4\pi}F\frac{\partial F}{\partial x} - \frac{\mu_p F}{\alpha}\frac{\partial j_n}{\partial F}\frac{\partial F}{\partial x}. \qquad (5.30)$$

The electron current density can be written

$$j_n = ne\mu_n F = -(\mu_n F/\alpha)(\partial j_n/\partial F)(\partial F/\partial x), \qquad (5.31)$$

and elimination of $F(\partial F/\partial x)$ from eqns (5.30) and (5.31) gives

$$\left\{\frac{\mu_n}{\alpha}j-\left(\frac{\mu_n}{\alpha}+\frac{\mu_p}{\alpha}\right)j_n\right\} = -\frac{\epsilon\mu_p}{4\pi}j_n. \qquad (5.32)$$

Using eqn (5.28) to separate the variables in (5.32), we get

$$\exp(-H/F)\,\mathrm{d}F = \frac{4\pi}{\epsilon\alpha_n}\left\{-\frac{\mu_n}{\mu_p}\frac{j}{j_n}\,\mathrm{d}j_n+\left(1+\frac{\mu_n}{\mu_p}\right)\,\mathrm{d}j_n\right\},$$
$$(5.33)$$

and integration of this equation will give a relation between current density and field strength to which the boundary conditions can be fitted.

The writing of eqn (5.33) in dimensionless form is greatly facilitated by the condition

$$\mu_n/\mu_p \gg 1, \qquad (5.34)$$

which we assume to hold in the remainder of this discussion. A dimensionless field strength $\mathscr{F}$ can be defined by

$$\mathscr{F} = F/H \qquad (5.35)$$

and a dimensionless current density J by

$$J = (4\pi\mu_n/\mu_p\epsilon\alpha_n H)j. \qquad (5.36)$$

Use of eqns (5.34), (5.35), and (5.36) in (5.33) gives

$$\exp(-1/\mathscr{F})\,\mathrm{d}\mathscr{F} = -(J/J_N)\,\mathrm{d}J_N+\mathrm{d}J_N, \qquad (5.37)$$

where J_N is related to j_n by the same constant that relates J to j. Integration of (5.37) can be carried out in terms of the exponential integral $E_1(z)$, which is defined by

$$E_1(z) = \int\limits_{z}^{\infty} \frac{\exp(-t)}{t}\,\mathrm{d}t \qquad (5.38)$$

and is readily available in tables, e.g. Abramowitz and Stegun (1964). The result is

$$\phi(\mathscr{F}) = -J\ln J_N+J_N+\mathrm{const}. \qquad (5.39)$$

where

$$\phi(\mathscr{F}) = (1/\mathscr{F})\exp(-1/\mathscr{F})-E_1(1/\mathscr{F}). \qquad (5.40)$$

The constant in eqn (5.39) can be evaluated using the boundary conditions. For our present purposes both contacts are blocking, but in order to accommodate the calculations of the following section it is convenient to fit the condition of an anode contact blocking for hole emission, viz

$$J = J_N \quad \text{when} \quad \mathscr{F} = \mathscr{F}_{\text{an}}, \qquad (5.41)$$

and leave the cathode boundary condition open for the time being. Using eqn (5.41) in eqn (5.39), we then get

$$J^{-1}\{\phi(\mathscr{F}) - \phi(\mathscr{F}_{\text{an}})\} = -\ln(J_N/J) + J_N/J - 1. \quad (5.42)$$

For a given total current J and anode field strength $\mathscr{F}_{\text{an}}$, this equation relates the electronic component of current to the field strength. The corresponding distance (measured from the anode) can be found from eqns (5.26) and (5.27), which combine to give

$$dx = -\frac{\mu_n}{\alpha_n} \frac{F}{\exp(-H/F)} \frac{dj_n}{j_n}. \qquad (5.43)$$

On introduction of a dimensionless distance by

$$X = (\alpha_n/\mu_n H)x, \qquad (5.44)$$

eqn (5.43) becomes

$$dX = -\{\mathscr{F}/\exp(-1/\mathscr{F})\} \, dJ_N/J_N. \qquad (5.45)$$

Equations (5.42) and (5.45) provide a means of calculating $\mathscr{F}$, J_N, and X for given J and $\mathscr{F}_{\text{an}}$.

The computation procedure is as follows.

(1) For assumed values of $\mathscr{F}_{\text{an}}$ and J compute J_N/J as a function of $\mathscr{F}$ from eqn (5.42).

(2) Compute X as a function of $\mathscr{F}$ from eqn (5.45) using the results of the first part of the computation.

(3) Compute the mean field

$$\overline{\mathscr{F}} = (1/X) \int_0^x \mathscr{F} \, dx. \qquad (5.46)$$

(4) For selected values of X the quantity J_N can then be plotted as a function of $\mathscr{F}$, and $\mathscr{F}$ as a function of $\overline{\mathscr{F}}$. If the selected values of X are regarded as the dimensionless thickness

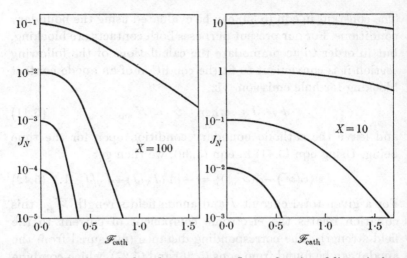

FIG. 5.4. The dimensionless electron current J_N as a function of dimensionless cathode field strength $\mathscr{F}_{\text{cath}}$ for various values of dimensionless total current J (marked on the ordinate). (a) $X = 100$, (b) $X = 10$. For $X = 1$ and $X = 10^{-1}$ it was found that $J_N \simeq J$ over the range considered.

of the dielectric, the corresponding values of $\mathscr{F}$ are then the dimensionless cathode field strength $\mathscr{F}_{\text{cath}}$. These plots are shown in Figs. 5.4 and 5.5.

For the case of double ejection the appropriate cathode boundary condition is

$$J_N/J = 0 \quad \text{when} \quad \mathscr{F} = \mathscr{F}_{\text{cath}} \qquad (5.47)$$

A study of Fig. 5.4 reveals that double ejection could occur only for the thickest dielectric samples. Using a slightly different but nevertheless equivalent approach to the computation, O'Dwyer (1968) calculated the double-ejection current characteristics and his results are shown in Fig. 5.6.

To estimate orders of magnitude we assume

$$\left. \begin{array}{l} H = 10 \text{ MV cm}^{-1} \\ \alpha_n = 10^8 \text{ s}^{-1} \\ \mu_n = 10^{-3} \text{ cm}^2 \text{ V}^{-1} \text{ s}^{-1} \\ \mu_n/\mu_p = 10^6 \\ \epsilon = 3 \end{array} \right\}. \qquad (5.48)$$

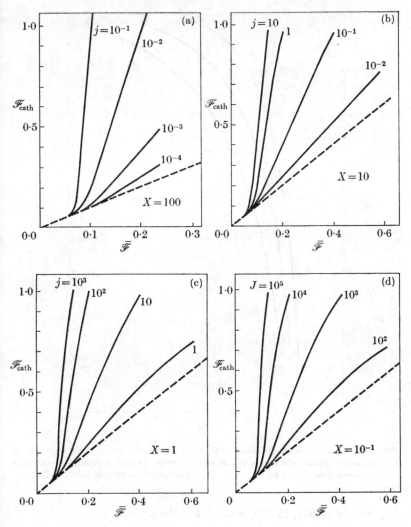

FIG. 5.5. The dimensionless cathode field strength $\mathscr{F}_{\text{cath}}$ as a function of dimensionless mean field strength $\bar{\bar{\mathscr{F}}}$ for various dimensionless total currents J. (a) $X = 100$, (b) $X = 10$, (c) $X = 1$, (d) $X = 10^{-1}$.

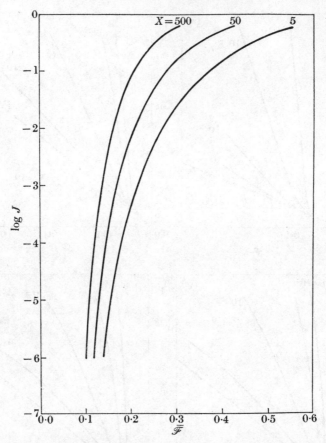

FIG. 5.6. Current–mean-field-strength characteristics for double ejection in dimensionless form for various values of thickness. The field-strength minimum is assumed adjacent to the cathode, corresponding to condition (5.34).

Equations (5.35), (5.36), and (5.44) then yield

$$\left.\begin{array}{l} j \simeq (5 \times 10^{-5}\,J)\ \text{A cm}^{-2} \\ x \simeq (10^{-4}\,X)\ \text{cm} \\ \bar{F} \simeq 10\bar{\mathscr{F}}\ \text{MV cm}^{-1} \end{array}\right\}. \tag{5.49}$$

These figures correspond to very small currents for large field strength in thick samples. It appears very unlikely that double-ejection currents could ever be observed; however it will be

shown below that a conduction mechanism may play a determining role in dielectric breakdown, even though other conduction currents completely blanket it in the pre-breakdown
region.

5.4. Contacts with field-dependent injection characteristics

If the mathematically simple boundary conditions used to
describe either ohmic or blocking contacts are abandoned in
favour of more realistic electrode conditions, the problem is
considerably more difficult. A complete dimensionless solution
is no longer possible, since the properties of both electrodes and
insulator now enter the problem.

In principle, one may assume any functional relation for the
injection characteristics (5.6), but in view of the difficulty of
the problem only fairly simple cases have so far been investigated. Usually one electrode is assumed to be blocking, say the
anode for hole emission, and eqn (5.6b) reduces to (5.5b). For
the cathode emission characteristics either a Fowler–Nordheim
or a Schottky-type mechanism is assumed, and eqn (5.6a) is
then written in the form of either (3.1) or (3.52) respectively.
More elaborate or modified versions of these equations could be
used, but the nature of the field dependence of the emission
current would remain the same.

In order that the problem should not simply reduce to an
electrode-determined one, the conduction characteristics of the
dielectric must be such that it can support a non-uniform space
charge with constant current. If the charge-carrier mobilities
are not field-dependent, this can be so only if collision ionization
occurs and the current is ambipolar. If single-carrier conduction
is considered, the effective mobility must be field-dependent, or
the assumptions of non-uniform space charge and constant
current become contradictory; the cases investigated have been
the Fröhlich amorphous dielectric and the Poole–Frenkel
dielectric, both of which lead to a field-dependent effective
mobility.

(a) Collision ionization in the dielectric

A theory of high-field conduction in a dielectric in which collision ionization by electrons occurs is given by O'Dwyer (1969a). It is assumed that the anode is completely blocking to hole emission, and that the cathode emits electrons in accordance with either the Fowler–Nordheim or the Schottky law. Since homogeneous boundary conditions are retained for the anode, it is possible to construct families of solutions in dimensionless form for matching to the inhomogeneous boundary condition at the cathode. The equations to be solved are

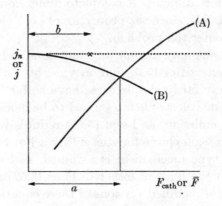

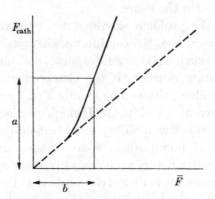

Fig. 5.7. Diagram illustrating a graphical method of fitting the cathode emission characteristic to the solutions of eqns (5.42) and (5.45).

identical with those of §5.3, and the computed solution shown
in Figs. 5.4 and 5.5 is in a form suitable for fitting the cathode
boundary condition.

To achieve this, the dimensionless formulation can no longer
be retained, and it is necessary to assume values for H, α_n,
μ_n, and ϵ, all of which are parameters of the dielectric. This
having been done, a simple graphical method for determining
the current–mean-field-strength characteristics can be followed.
The method is explained with reference to Fig. 5.7. Let the
curve labelled (A) be the assumed emitting characteristic of the
cathode. Transfer the solution j_n against F_{cath} (for some
assumed thickness and total current) from Fig. 5.4; this is the
curve labelled (B). The total current j is then represented by the
dotted horizontal line. The intersection of curves (A) and (B)
gives the values of j_n at the cathode and F_{cath} appropriate to the
particular injecting characteristic, thickness, and total current.
Then from Fig. 5.5 (with dimensions now entered) the corresponding value of the mean field strength $\bar{F}$ can be found. If
this is plotted as abscissa on the dotted line of the assumed total
current, one then has a point on the current–mean-field-
strength characteristic.

This procedure was carried out for two cases shown in Fig.
5.8 using the dielectric parameters quoted in (5.48). Figure
5.8(a) corresponds to a cathode emitting with a Fowler–
Nordheim characteristic and a work function $\phi = 1 \cdot 25$ eV;
Fig. 5.8(b) corresponds to Schottky emission at room temperature. (From a single characteristic the work function and
pre-exponential factor are not independently determined.) In
both cases, the positive space charge in the dielectric causes
little variation from the cathode emission characteristic up to a
critical current (dependent on the dielectric thickness); for
higher current densities, the steady-state solution is unstable
and corresponds to a current-controlled negative resistance.

(b) Fröhlich amorphous dielectric

The steady-state current–voltage characteristics corresponding to Fowler–Nordheim or Schottky emission into a

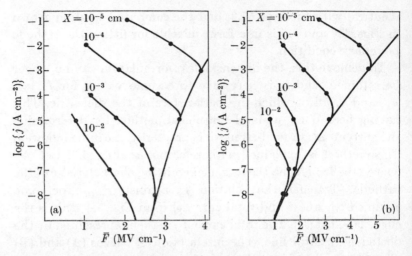

FIG. 5.8. Current–mean-field-strength characteristics for two special cases described in the text. The dielectric parameters are given by (5.48). (a) Fowler–Nordheim cathode emission with $\phi = 1\cdot25$ eV. (b) Schottky cathode emission at room temperature.

Fröhlich amorphous dielectric (whose field-dependent conduction characteristics were derived in §4.3(a)) were calculated by O'Dwyer (1966). In this problem only electron conduction is considered; the anode is assumed blocking to hole emission and there is no collision ionization.

Continuity of current in the body of the dielectric requires

$$j = \sigma F = \text{const.} \tag{5.50}$$

where both F and σ are functions of position. The variation of F implies a space charge in the dielectric, and the conductivity will then be variable not only because of the variation of the field, but also because of the space charge. Equation (4.68) will therefore require modification to take account of this space charge. Let n_0 be the electron density over all traps and conduction levels for the space-charge-free condition, and let n be the electron space-charge density that actually occurs. Then eqn (4.68) is modified to become

$$\sigma(F) = \sigma(0)\left(1 + \frac{n}{n_0}\right)\exp\left(\frac{W}{\Delta V}\frac{F^2}{F_c^2}\right), \tag{5.51}$$

where F_c is the critical field strength given by (4.65). Combining eqns (5.50) and (5.51), we have

$$n = n_0\left\{\frac{j}{\sigma(0)F}\exp\left(-\frac{W}{\Delta V}\frac{F^2}{F_c^2}\right) - 1\right\},\qquad(5.52)$$

giving the space-charge density as a function of the current and the variable field strength. Poisson's equation can be written

$$\partial F/\partial x = 4\pi n e/\epsilon,\qquad(5.53)$$

where in this problem the distance x is measured from the cathode. Substituting eqn (5.52) in eqn (5.53) and separating the variables, we obtain

$$x = \frac{\epsilon}{4\pi n_0 e}\int_{F_{cath}}^{F}\frac{\mathrm{d}F}{\{j/\sigma(0)F\}\exp(-WF^2/\Delta V F_c^2)-1}.\qquad(5.54)$$

This expression gives the distance from the cathode to the variable point, and forms a suitable basis for computing the solution.

The computation procedure is as follows.

(1) Select a value of the current density j; F_{cath} is then determined from either (3.1) or (3.52), corresponding to Fowler–Nordheim and Schottky emission respectively.

(2) For assumed parameters of the dielectric, eqn (5.54) allows a computation of $F = F(x, j)$.

(3) The voltage can then be computed from

$$\int_0^L F\,\mathrm{d}x = V(j)$$

to yield the desired result.

These computations were carried out by O'Dwyer (1966) for dielectric parameters given by

$$\left.\begin{aligned}\epsilon &= 5\\ n_0 &= 10^{18}\ \mathrm{cm}^{-3}\\ \sigma_0 &= 10^{-12}\ \Omega^{-1}\ \mathrm{cm}^{-1}\\ W/\Delta V &= 5\\ F_c &= 10\ \mathrm{MV}\ \mathrm{cm}^{-1}\end{aligned}\right\}\qquad(5.55)$$

and for three values of the cathode work function,

$$\phi = 1\cdot5,\ 1\cdot0,\quad \text{and}\quad 0\cdot85\ \text{eV}. \tag{5.56}$$

The first two values in (5.56) were used in conjunction with eqn (3.1) to calculate Fowler–Nordheim emission, and the last value was used in eqn (3.52) to calculate room temperature Schottky emission. The results of the computations are plotted on Schottky diagrams in Fig. 5.9; all cases show considerable

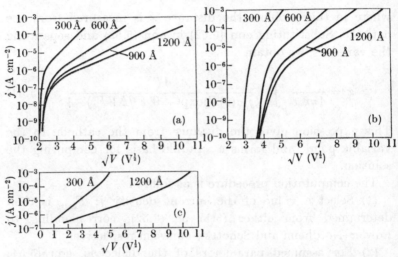

FIG. 5.9. Schottky plots of computed current–voltage characteristics from various cathodes into a Fröhlich amorphous dielectric for the dielectric parameters given in (5.55). (a) Fowler–Nordheim emission, $\phi = 1\cdot5$ eV. (b) Fowler–Nordheim emission, $\phi = 1\cdot0$ eV. (c) Schottky emission, $\phi = 0\cdot85$ eV.

regions in which the characteristic is linear regardless of the nature of the cathode emission.

It is particularly intriguing that a Fowler–Nordheim cathode should combine with a Fröhlich amorphous dielectric to yield a linear region on a Schottky diagram; in fact it can be true over only a limited range of thickness since a very thin film would exhibit a tunnelling characteristic, and a very thick film would show a conductivity characterized by eqn (4.68).

Lilly, Lowitz, and Schug (1968) investigated the parameters required to fit this theory to conduction data on Mylar films

taken by Lilly and McDowell (1968). The dielectric constant
was taken to be the high-frequency value 2·44, $W/\Delta V$ and F_c
were assumed as given in (5.55), and the remaining parameters
were chosen to fit the data. The results of their analysis are
shown in Fig. 5.10. It is apparent that the parameters required

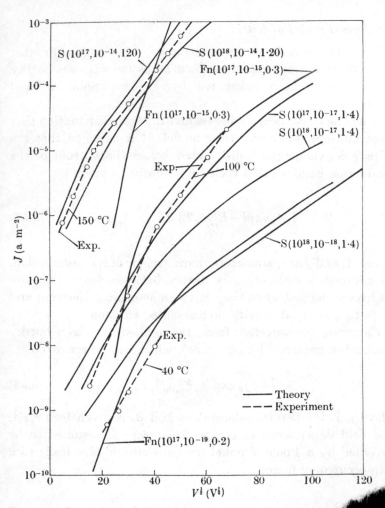

FIG. 5.10. Comparison of the theory of §5.4(b) with experiment
1-mil Mylar (1 mil = $\frac{1}{1000}$ inch). The 'Fn' stands for Fowler–No
the 'S' for Schottky cathodes. The figures in brackets are $(n_0(\text{m}^{-1}),\ \phi(\text{eV}))$.

.or best fit are of the correct order of magnitude, and in the case of the Schottky cathode a reasonable value of the work function is also obtained. The work function required for good fit of the data with a Fowler–Nordheim cathode is too low to be seriously considered.

(c) Poole–Frenkel dielectric

The current–voltage characteristics of a Poole–Frenkel dielectric containing a distribution of traps and with a Schottky cathode have been calculated by Pulfrey, Shousha, and Young (1970).

It will be convenient to discuss first the trap distribution that these authors assume for their model. It is assumed that the traps are exponentially distributed below the bottom of the conduction band, so that their total density is given by

$$N_t = \int_0^\infty A_t \exp(-E/\gamma k_0 T)\, dE = A_t \gamma k_0 T, \qquad (5.57)$$

where A_t and γ are parameters characteristic of the distribution. In addition a distinction is drawn between traps that are positively charged when they have no associated electron, and traps that are neutral with no associated electron.

Electrons are injected from the cathode by a Schottky mechanism governed by eqn (3.52), which can be rewritten

$$j = j_0 \exp(\beta_S F_{\text{cath}}^{\frac{1}{2}}/k_0 T), \qquad (5.58)$$

where j_0 is temperature-dependent and β_S is given by $(e^3/\epsilon)^{\frac{1}{2}}$. The field dependence of escape from traps is assumed to be governed by a Poole–Frenkel mechanism, and this leads to a trap-occupancy factor

$$\Big)^2 F(x)\exp\{(\beta_{\text{PF}}F(x)-\beta_S F_{\text{cath}}-E)/k_0 T\}\Big]^{-1} \quad (5.59)$$

where $f_D(x, E)$ depends on distance and energy below the conduction levels. It is also assumed that

$$\beta_{PF} = 2\beta_S. \tag{5.60}$$

It has been pointed out in §4.3(b) that this assumption is appropriate to highly compensated materials. This corresponds to the case in which traps are neutral when they have no associated electron, and we shall therefore confine ourselves to this case.

The electric-field distribution in the dielectric can then be determined from Poisson's equation; the results of Pulfrey *et al.* (1970) are shown in Fig. 5.11; the trapped-electron density

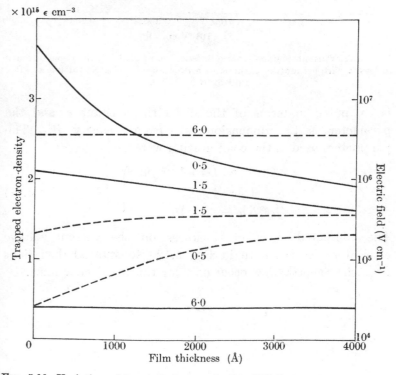

FIG. 5.11. Variation of trapped electron density (full line) and electric field (dashed line) with position in the dielectric film for the case of trapping centres which are neutral when empty, and for various values of dimensionless current density (marked on the curves) (after Pulfrey *et al.* 1970).

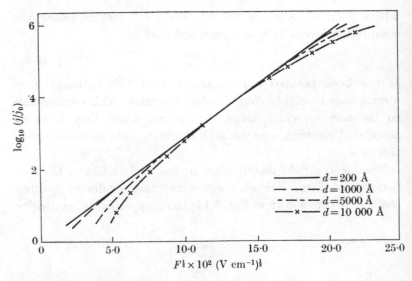

FIG. 5.12. Current–voltage characteristics computed for a Poole–Frenkel dielectric with a Schottky cathode as a function of dielectric thickness (after Pulfrey *et al.* 1970).

is computed in terms of the dielectric constant ϵ, and the parameter is the dimensionless current density j/j_0. The parameters used in the computation were

$$\left.\begin{array}{l} N_t = 1{\cdot}4 \times 10^{16} \text{ cm}^{-3} \\ C_2 = 10^5 \\ j_0\beta_S^2/e\mu = 5{\cdot}5 \times 10^9 \epsilon \end{array}\right\}. \tag{5.61}$$

The computed effect of thickness on the current–voltage characteristics is shown in Fig. 5.12. Substantial deviations from the Schottky law occur only for thick dielectric films.

THERMAL BREAKDOWN

6.1. General principles of thermal breakdown

IT has been frequently said that all dielectric breakdown is ultimately in some manner thermal, but the name 'thermal breakdown' has been reserved for the type of breakdown that can be adequately described in terms of the thermal properties of the dielectric and the pre-breakdown electrical conductivity. It is governed by eqn (1.7), which we restate for convenience:

$$C_V \frac{\partial T}{\partial t} - \text{div}(\kappa \, \text{grad} \, T) = \sigma F^2, \tag{6.1}$$

where C_V is the specific heat per unit volume, and σ and κ are the electrical and thermal conductivities respectively. (For an a.c. field the d.c. conductivity must be replaced by the appropriate loss factor.) Since we shall deal only with cases of no charge accumulation, the equation of current continuity is

$$\text{div}(\sigma \mathbf{F}) = 0. \tag{6.2}$$

Calculation of a critical thermal situation then involves a solution of eqns (6.1) and (6.2) of the form

$$T = T(t, \mathbf{r}), \tag{6.3}$$

in which $\mathbf{r}$ is an arbitrary point in the dielectric. For a given manner of application of the field

$$\mathbf{F} = \mathbf{F}(t, \mathbf{r}) \tag{6.4}$$

and this relationship depends on the electrode configuration. For the solution of eqn (6.1) the dependence of the electrical and thermal conductivities on the parameters is also required, viz.

$$\sigma = \sigma(F, T) \tag{6.5}$$

and

$$\kappa = \kappa(T). \tag{6.6}$$

Although the thermal conductivity is frequently slightly temperature-dependent, it is usual to treat it as a constant in theories of thermal breakdown; the temperature and field-strength dependence of electrical conductivity are often explicitly considered. In principle there are therefore as many thermal-breakdown theories as there are electrode geometries to determine the form of (6.4), and theories of conductivity to determine (6.5).

The criterion for breakdown is then that T exceeds some assigned critical value, the exact nature of which turns out to be unimportant in most calculations. The time and position for this critical temperature being known, it is often convenient to determine a critical field strength from (6.4) as specifying thermal instability under these conditions. The last step is by no means essential, and other criteria may just as validly be used, e.g. the total inter-electrode voltage difference may be calculated as specifying critical conditions.

Because of the difficulties which the functional forms (6.4) and (6.5) may introduce, this solution cannot be carried out for the general case. Three main attacks can be made on the problem: determination of steady-state thermal breakdown, which involves ignoring the time derivative in (6.1); determination of impulse thermal breakdown by ignoring the heat-conduction term in (6.1); and numerical or approximate solutions of (6.1) for simple cases. We proceed to investigate these main lines.

6.2. Steady-state thermal breakdown (one dimension)

A detailed treatment of steady-state thermal breakdown in the one-dimensional case was first given by Fock (1927), and later discussed and reviewed by other authors (cf. Moon 1931 and Whitehead 1951).

Throughout this section we shall consider a slab or homogeneous dielectric of thickness d and of large area, the electrodes being so thin as to constitute no thermal impediment between the dielectric and the ambient medium. Furthermore, we shall consider only the direct-current case with the voltage raised slowly, so that a steady state always obtains. If the z direction is taken perpendicular to the electrode surfaces, then all heat flow will be in the z direction, and z will be the only coordinate variable entering the equations. With these restrictions, eqn (6.1) reduces to

$$\frac{\partial}{\partial z}\left(\kappa\,\frac{\partial T}{\partial z}\right)+\sigma\left(\frac{\partial V}{\partial z}\right)^{2}=0. \qquad (6.7)$$

The equation of electrical continuity takes the form

$$\frac{\partial}{\partial z}\left(\sigma\,\frac{\partial V}{\partial z}\right)=0,$$

which can be integrated at once to give

$$\sigma\partial V/\partial z=-j. \qquad (6.8)$$

The equations (6.7) and (6.8) are sufficiently difficult to solve (with prescribed boundary conditions) to render the prospect of solving more complex situations unprofitable, all the more so since the one-dimensional solution will be approximately correct if the ratio of dielectric thickness to electrode dimension is small.

(a) The maximum thermal voltage

It is of considerable interest that, in one dimension, a maximum thermal voltage can be derived as a certain limiting case (cf. Whitehead (1951)). The limit concerns the situation in which an infinite dielectric slab of arbitrary thickness is constrained to the ambient temperature at its bounding electrode surfaces by a cooling system of sufficient capacity. The thermal insulation provided by the dielectric itself is then the limiting factor on the dissipation of energy—in fact for a sufficiently thick slab of dielectric it is clear that the surface conditions will have little effect. (In this discussion we do not consider the case of electrodes that are perfect thermal insulators, since clearly there will then be no possible stable situation, no matter how low the applied voltage.) The notation for the problem is shown in Fig. 6.1, in which z is measured

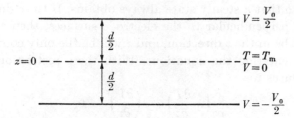

FIG. 6.1. Diagram of notation for infinite dielectric slab.

from the central plane at which the temperature is by symmetry a maximum T_m. The temperature of the dielectric at the surface is taken in the general case as T_1, and the temperature of the ambient medium T_0.

The equations (6.7) and (6.8) combine to give

$$\frac{\partial}{\partial z}\left(\kappa\,\frac{\partial T}{\partial z}\right) - j\,\frac{\partial V}{\partial z} = 0. \tag{6.9}$$

Integrating from the centre to a variable point, we have

$$jV = \int_0^z \frac{\partial}{\partial z}\left(\kappa\,\frac{\partial T}{\partial z}\right)\,\mathrm{d}z = \kappa\,\frac{\partial T}{\partial z}, \tag{6.10}$$

since $\partial T/\partial z = 0$ at the central plane. Substitution for j from (6.8) in (6.10) gives

$$V = -\frac{1}{\sigma}\,\frac{\partial z}{\partial V}\,\kappa\,\frac{\partial T}{\partial z},$$

which becomes on separation of the variables

$$V\,\mathrm{d}V = -(\kappa/\sigma)\,\mathrm{d}T. \tag{6.11}$$

Integrating from the centre to a variable point and reversing the limits we get

$$V^2 = 2\int_T^{T_m} (\kappa/\sigma)\,\mathrm{d}T. \tag{6.12}$$

If the integration proceeds to the electrode, $V = V_0/2$ and $T = T_1$, so that

$$V_0^2 = 8\int_{T_1}^{T_m} (\kappa/\sigma)\,\mathrm{d}T. \tag{6.13a}$$

For the limiting case discussed in this section, $T_1 = T_0$ (most efficient edge cooling) and the critical voltage is given by

$$V_{0c}^2 = 8 \int_{T_0}^{T_{mc}} (\kappa/\sigma) \, \mathrm{d}T, \qquad (6.13b)$$

where T_{mc} is the critical value of the maximum temperature. The electrical-conductivity laws with which we are concerned are such that eqn (6.13b) is insensitive to the precise value of T_{mc} and in many cases it may be taken as infinite without serious error. It is clear that the result (6.13) stems from the unidimensional conditions, and that, if the electrode arrangements allowed lateral heat flow, the thermal-breakdown voltage could be raised indefinitely.

It should be noted that the separation of variables carried out in arriving at eqn (6.11) cannot be accomplished if the electrical conductivity is field-strength-dependent. However, for many dielectrics the conductivity can be written in the form of eqn (1.2) if the field strength is not too high. Substitution of (1.2) and $\kappa = \kappa_0$ in eqn (3.58b) gives

$$V_{0c}^2 = 8 \int_{T_0}^{T_{mc}} \frac{\kappa_0}{\sigma_0} \exp(\phi/k_0 T) \, \mathrm{d}T,$$

from which we find

$$V_{0c} \simeq \left(\frac{8\kappa_0 k_0 T_0^2}{\sigma_0 \phi}\right)^{\frac{1}{2}} \exp(\phi/2k_0 T_0), \qquad (6.14)$$

the approximation in (6.14) being contingent on the inequalities $\phi \gg k_0 T$ for all temperatures considered and $T_{mc} > T_0$.

Thermal breakdown was observed in thick single-crystal plates of NaCl by Inge, Semenoff, and Walther (1925). Using a.c. voltages of slowly increasing amplitude, they found evidence of thermal breakdown in NaCl at temperatures greater than about 220 °C. A typical set of results for breakdown voltage as a function of thickness is shown in Fig. 6.2, and the constant critical voltage for thick plates is clearly evident. At 700 °C the conductivity of NaCl should be intrinsic

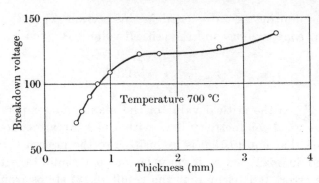

FIG. 6.2. Breakdown voltage of NaCl slabs of varying thickness at 700 °C with voltage rise time of the order of minutes (after Inge *et al.* 1925).

and given by eqn (2.9), which is of the form (1.2) required for the validity of (6.14). Although there is some uncertainty about the parameters to be used, an order-of-magnitude check on the theory can be made by using the following values:

$$\kappa_0 = 4 \times 10^{-3} \text{ cal cm}^{-1} \text{ s}^{-1} \text{ °C}^{-1} \text{ at 700 °C} \quad \text{(Inge } et\, al.\ 1925);$$

$$\sigma = 10^{-4} \ \Omega^{-1} \text{ cm}^{-1} \text{ at 700 °C (Seitz 1940)},$$

$\phi = 1 \cdot 85$ eV (at this temperature the ionic conduction is intrinsic) (Jacobs and Tompkins 1952).

Substitution in (6.14) yields $V_{0c} = 240$ V which is reasonably satisfactory agreement in view of the parameter uncertainties, and the lack of knowledge of the thermal resistance of the electrode system.

(b) *The general solution for field-independent conductivity*

We consider now the case in which thermal conduction with non-ideal boundaries is treated in detail, but the electrical conductivity is not field-strength-dependent. We assume that there exists a constant external thermal conductivity λ such that the heat lost by the dielectric surface at temperature T_1 to the ambient at temperature T_0 is given by $\lambda(T_1 - T_0)$ per unit area per unit time. If a steady state exists, this implies that

$$\tfrac{1}{2} j V_0 = \lambda (T_1 - T_0), \tag{6.15}$$

since half of the heat generated by the current will be lost through each electrode surface. Substitution of (6.15) in (6.9) yields

$$\frac{\partial}{\partial z}\left(\kappa\,\frac{\partial T}{\partial z}\right) - \frac{2\lambda(T_1 - T_0)}{V_0}\,\frac{\partial V}{\partial z} = 0, \tag{6.16}$$

which gives on integration

$$\kappa\,\frac{\partial T}{\partial z} - \frac{2\lambda(T_1 - T_0)}{V_0}\,V = 0. \tag{6.17}$$

The constant of integration is zero since $V = 0$ when $\partial T/\partial z = 0$ at the central plane. Substituting (6.12) in (6.17) and integrating from the centre to the edge, we get

$$\frac{\lambda(T_1 - T_0)d}{V_0} = \int_{T_1}^{T_\mathrm{m}} \frac{\kappa\,\mathrm{d}T}{\left(\int_{T}^{T_\mathrm{m}} \frac{2\kappa}{\sigma}\,\mathrm{d}T\right)^{\frac{1}{2}}}. \tag{6.18}$$

Equations (6.13a) and (6.18) determine the thermal breakdown.

In general the principle of solution of eqns (6.13a) and (6.18) is to eliminate T_1 and express T_m as a function of T_0 for some given V_0. The result of such a calculation is shown schematically in Fig. 6.3, from which it appears that there are two solutions

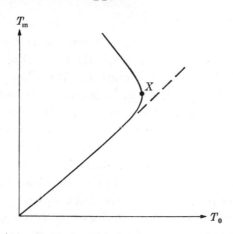

FIG. 6.3. Diagram to illustrate the determination of critical conditions for steady-state thermal breakdown.

T_m (the lower stable and the upper unstable) for T_0 sufficiently low. At the point X critical conditions apply, and for higher ambient temperatures thermal breakdown occurs at the given V_0. This argument can of course be phrased equivalently to yield a critical V_0 at a given T_0, since it is a combination of both that is critical.

We shall examine the case for which the thermal conductivity is constant and the electrical conductivity is given by eqn (1.2). With this restriction, eqns (6.13a) and (6.18) can be written in terms of dimensionless quantities as follows.

$$v^2 = \int_{x_m}^{x_1} \frac{\exp x}{x^2}\, \mathrm{d}x \tag{6.19}$$

and

$$\frac{c}{v}\left(\frac{1}{x_1} - \frac{1}{x_0}\right) = \int_{x_m}^{x_1} \frac{\mathrm{d}x}{x^2 \left\{ \int_{x_m}^{x} \frac{\exp x}{x^2}\, \mathrm{d}x \right\}^{\frac{1}{2}}}, \tag{6.20}$$

where we have introduced the dimensionless quantities

$$v = V_0 (k_0 \sigma_0 / 8\phi\kappa_0)^{\frac{1}{2}}, \tag{6.21}$$

$$c = \lambda d / 2\kappa_0, \tag{6.22}$$

$$x = \phi / k_0 T \tag{6.23}$$

(cf. Moon 1931), which are proportional to voltage, thickness, and reciprocal temperature respectively. (Subscripts on x in (6.23) imply the same subscript on T.) The procedure for solution is now to eliminate x_1 from eqns (6.19) and (6.20), use the critical condition to determine x_m in terms of x_0, and eventually v as a function of c and x_0. This greatly simplifies the presentation of the solution in a form suitable for all dielectrics that obey the conductivity law (1.2). Results of the numerical solution of eqns (6.19) and (6.20) are shown in Fig. 6.4 which was prepared from the Fock tables given by Moon (1931). The quantity v is plotted as a function of c (both on logarithmic scales) for various values of x_0 considered as a parameter.

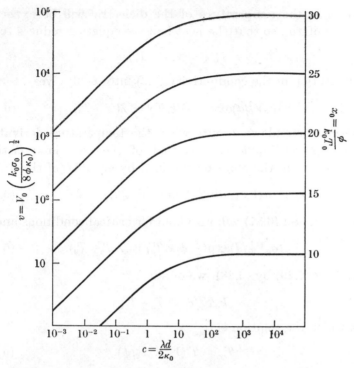

FIG. 6.4. Chart showing the results of the numerical solution of eqns (6.19) and (6.20) for steady-state thermal breakdown.

It is clear from Fig. 6.4 that simple limiting forms of solution should be possible for values of c either very much larger or very much smaller than unity. Consider first the case for which

$$\lambda d \gg 2\kappa_0. \qquad (6.24)$$

This gives an exact meaning to what has simply been referred to as a thick slab in § 6.2(a), and the appropriate approximate solution (6.14) has already been derived.

The condition for a thin slab is

$$\lambda d \ll 2\kappa_0, \qquad (6.25)$$

and it is somewhat laborious to find the corresponding approximate formula for breakdown from eqns (6.19) and (6.20); the result is found more easily from the following argument. For a

thin slab, the temperature of the dielectric will be expected
to be uniform, so that the heat balance equation reduces to

$$jV_0 = \lambda(T - T_0). \tag{6.26}$$

Substitution of the conductivity (1.2) into (6.26) gives

$$(\sigma_0 V_0^2/d)\exp(-\phi/k_0 T) = \lambda(T - T_0). \tag{6.27}$$

Thermal breakdown occurs when the temperature derivative
of the power input exceeds that of the heat loss; critical
conditions will therefore be specified by equality of the derivatives

$$(\sigma_0 V_{0c}^2\phi/dk_0 T_c^2)\exp(-\phi/k_0 T_c) = \lambda. \tag{6.28}$$

However, eqn (6.27) will also hold for critical conditions and it
becomes

$$(\sigma_0 V_{0c}^2/d)\exp(-\phi/k_0 T_c) = \lambda(T_c - T_0). \tag{6.29}$$

Dividing (6.29) by (6.28), we obtain

$$k_0 T_c^2/\phi = T_c - T_0,$$

for which the solution is

$$T_c = T_0(1 + k_0 T_0/\phi). \tag{6.30}$$

Substitution of (6.30) into (6.29) gives for the critical voltage

$$V_{0c}^2 = \frac{\lambda k_0 T_0^2 d}{\sigma_0 \phi e} \exp(\phi/k_0 T_0), \tag{6.31}$$

where e is the natural base of logarithms. Equation (6.31) is
the thin-film approximation to the solutions of eqns (6.19) and
(6.20) shown in Fig. 6.4. It is interesting that eqn (6.31) gives
the critical voltage proportional to the square root of the
thickness, so that critical conditions in a thin slab are specified
by neither a voltage nor a field strength. The experimental
results of Inge et al. (1925) shown in Fig. 6.2 depict a breakdown
voltage varying approximately with the square root of thickness
for NaCl slabs thinner than 1 mm.

Essentially the same result was found by Whitehead and
Nethercot (1935) (cf. Whitehead 1951) and by Klein and Gafni
(1966) (cf. Klein 1969). The latter used a conductivity relation

of the form
$$\sigma = \sigma_0 \exp(aT) \tag{6.32}$$

from which the thin-film approximation is readily found to be

$$V_{0c}^2 = \frac{\lambda d}{\epsilon a \sigma_0} \exp(-aT_0) \tag{6.33}$$

in place of (6.31).

(c) *Thermal breakdown in thin films with field-dependent conductivity*

If field-strength dependence is formally included in relation (6.5) for the electrical conductivity, general equations of the type (6.19) and (6.20) become impossibly complicated, and no solution has been given. It is however possible to include an explicit field-strength dependence in the electrical conductivity if the thermal-conduction processes are treated in the simplified manner of eqn (6.26) appropriate to thin films. This was done by Whitehead and Nethercot (1935), Klein and Gafni (1966), and Sze (1967).

A simpler analysis results if the temperature and field-strength dependence of the conductivity are separable. Klein and Gafni (1966) proceeded in this way by assuming

$$\sigma = \sigma_0 \exp(aT + bF), \tag{6.34}$$

where σ_0, a, and b are material constants. Substitution of (6.34) into (6.26) gives

$$\sigma_0 F^2 d \exp(aT + bF) = \lambda(T - T_0). \tag{6.35}$$

One finds the critical conditions in exactly the same way as in the case of field-independent conductivity (eqns (6.26)–(6.31)). The analysis yields

$$F_c \simeq \frac{1}{b} \left\{ \ln\left(\frac{\lambda}{a \sigma_0 \epsilon d F_c^2} \right) - aT \right\}. \tag{6.36}$$

This is not an explicit expression for F_c, but it is a form suitable for calculations by successive approximations; the zero-order approximation is obtained by taking F_c as constant in the logarithmic term. It is noteworthy that the rather strong thickness dependence of the field-independent case (eqn (6.33)) has become a weak dependence in eqn (6.36). In addition, the temperature dependence has become linear rather than exponential. This linear dependence was found by Klein and Lizak, whose results for the breakdown strength of silicon oxide capacitors as a function of temperature are shown in Fig. 6.5. The calculated points on Fig. 6.5 were found by the

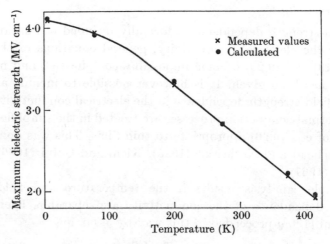

FIG. 6.5. Dielectric breakdown strength of silicon oxide capacitors as a function of temperature. The capacitors were 4100 Å thick and 0·06 cm² in area. (After Klein and Lizak 1966.)

authors from eqn (6.36) with a, b, and λ calculated as functions of temperature from the data of Fig. 4.18(a). This procedure is open to serious objection in that quantities treated as constant parameters in the calculation then become fairly strongly varying functions of the temperature (cf. Klein and Gafni (1966)). It should be more meaningful to compare the linear portion of Fig. 6.5, which can be expressed as

$$F^* = (4\cdot6 - 6\cdot5 \times 10^{-3}\ T)\ \text{MV cm}^{-1}, \qquad (6.37)$$

with the theoretical result (6.36). From the breakdown data we have
$$a/b = 6\cdot5 \times 10^{-3}\ \text{MV cm}^{-1}\ {}^{\circ}\text{C}^{-1},$$

while the results of Klein and Gafni (1966) taken from the conductivity data of Fig. 4.18(a) vary from

$$7 \cdot 8 \times 10^{-3} \text{ MV cm}^{-1} \, {}^{\circ}\text{C}^{-1} \quad \text{at} \quad 150 \text{ K}$$

to

$$21 \cdot 7 \times 10^{-3} \text{ MV cm}^{-1} \, {}^{\circ}\text{C}^{-1} \quad \text{at} \quad 350 \text{ K}.$$

The agreement is of the correct order of magnitude, but is certainly not so good as to demonstrate conclusively the thermal nature of the breakdown.

A more complex theoretical situation arises if one assumes a form for the electrical conductivity in which the field-strength and temperature dependences are not separable. Such a situation was investigated by Sze (1967), who considered a Poole–Frenkel type of relation for the conductivity,

$$\sigma = \sigma_0 \exp\{-(\phi - \beta F^{\frac{1}{2}})/k_0 T\}. \tag{6.38}$$

Substitution of (6.38) into (6.26) gives

$$\sigma F^2 d \exp\{-(\phi - \beta F^{\frac{1}{2}})/k_0 T\} = \lambda(T - T_0). \tag{6.39}$$

The calculations follow the familiar pattern to yield

$$F_c = \beta^{-2}(\phi - CT)^2, \tag{6.40}$$

where C is given by

$$C = k_0 \ln\left\{\frac{\sigma_0 F_c^2 d(\phi - \beta F_c^{\frac{1}{2}})}{\epsilon \lambda k_0 T_0^2}\right\} \tag{6.41}$$

and is assumed to be a slowly varying function of the variables which it contains. Experimental data on the breakdown strength of silicon nitride films found by Sze (1967) are shown in Fig. 6.6. In this case it is quite difficult to verify that the slope of the curve agrees with eqn (6.40), since the value of λ in (6.41) is not available from measurements other than the breakdown itself. However, the zero-temperature intercept should be given by ϕ/β^2, which is available from the conductivity measurements. The zero-temperature intercept from breakdown data is $11 \cdot 6$ MV cm^{-1}; using the data quoted by Sze (1967) for ϕ and β, one finds for ϕ/β^2 a value (16 ± 7) MV cm^{-1} which is of the correct order of magnitude.

The most doubtful point in making a definitive judgement

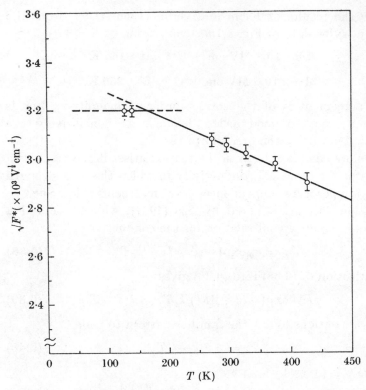

FIG. 6.6. Temperature dependence of the breakdown strength of silicon nitride films 1000 Å thick and $1 \cdot 6 \times 10^{-4}$ cm² in area (after Sze 1967).

regarding the nature of the breakdown process in thin films of silicon oxide and silicon nitride under the circumstances described above is that the value of the external thermal conductivity is available only from the breakdown measurements using the thermal-breakdown theory. An independent determination of λ would put the postulate of thermal breakdown on a much more secure footing, since other available checks agree in order of magnitude.

(d) Thermal breakdown in thick films with field-dependent conductivity

Chou and Brooks (1970) considered the problem of introducing a field-dependent conductivity into the calculations of

§6.2(a). The form chosen for the conductivity was

$$\sigma(T,\, F) = \sigma_0 \exp\{-(\phi - \beta F)/k_0 T\}, \qquad (6.42)$$

which is yet another functional form in addition to (6.34) and (6.38). The choice of (6.42) is justified by data such as those shown in Fig. 4.1(a). The separation of field-strength and temperature variables that was achieved in eqn (6.11) cannot now be accomplished because of the form of (6.42). Numerical integration was carried out by Chou and Brooks (1970) using data for NaCl taken from Hanscomb *et al.* (1966); the data used were

$$\left.\begin{array}{l} \kappa = 0\cdot05 \text{ J cm}^{-1}\text{ K}^{-1}\text{ s}^{-1} \\ \phi = k_0\, 12\,000 \text{ K} \\ \beta = k_0\, 614 \text{ K MV}^{-1}\text{ cm} \\ \sigma_0 = 71\cdot8\ \Omega^{-1}\text{ cm}^{-1} \\ T_0 = 250\ ^\circ\text{C} \end{array}\right\}, \qquad (6.43)$$

for which the computed curve shown in Fig. 6.7 was obtained. For the sake of comparison the result of using the data (6.43) in eqn (6.14) is also shown, and it is clear that inclusion of the field dependence in the conductivity is important if the sample thickness is less that 0·1 mm. The question of whether the thick-slab or thin-slab thermal-breakdown theory applies is of course decided by eqns (6.24) and (6.25), provided that the slab is an order of magnitude thicker than the phonon mean free path; it is always thick if λ is sufficiently large.

6.3. Impulse thermal breakdown

The opposite limiting case to that of a steady state is one in which the build-up of the electric field occurs so quickly that thermal-conduction processes play a negligible role. The fundamental equation (6.1) then becomes

$$C_V\, \mathrm{d}T/\mathrm{d}t = \sigma F^2, \qquad (6.44)$$

which determines the critical conditions for impulse thermal breakdown.

The nature of the critical conditions determined from eqn (6.44) differs somewhat from those derived from eqn (6.7). In

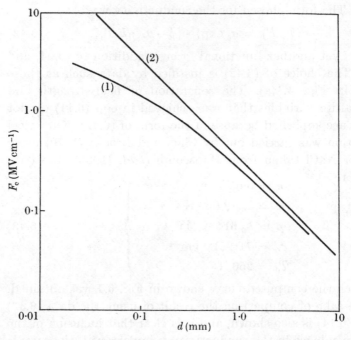

FIG. 6.7. Calculated thermal critical field strength of NaCl at 250 °C as a function of thickness. (1) Computed by Chou and Brooks (1970) using the high-field conductivity data of Hanscomb *et al.* (1966). (2) Plot of eqn (6.14) using low-field conductivity data.

the latter case, the electrode geometry determines the field distribution, and, in conjunction with the thermal properties of the electrodes, the steady-state heat-flow pattern. However, for impulse thermal breakdown the electrode geometry determines only the field-strength distribution to be used in eqn (6.44), which does not require spatial boundary conditions for solution. If the time dependence of the applied field pulse is known,

$$F = F(t), \qquad\qquad (6.45)$$

then an equation of the form (6.5) for the electrical conductivity can be used in conjunction with (6.45) to integrate (6.44). This process yields the combination of critical field strength F_c and critical time t_c required in order that T should exceed T_m.

This integration can be carried out in an approximate way if the electrical conductivity follows the law (1.2) and if (6.45) is any simple function. Considering the case of a field-strength pulse increasing linearly with time, i.e.

$$F = (F_c/t_c)t \qquad (6.46)$$

(cf. Vermeer 1956b, O'Dwyer 1960), we have, using (1.2) and (6.46) in (6.44),

$$\frac{dT}{dF} = \frac{t_c}{F_c} \frac{\sigma_0 F^2}{C_V} \exp(-\phi/k_0 T),$$

which after variable separation yields

$$\int_0^{T_m} \exp(\phi/k_0 T) \, dT = \frac{t_c}{F_c} \frac{\sigma_0}{C_V} \int_0^{F_c} F^2 \, dF. \qquad (6.47)$$

In the case $\phi \gg k_0 T$ and $T_m > T_0$, the integration of eqn (6.47) can be performed approximately to give

$$F_c \simeq \left\{ \frac{3C_V k_0 T^2}{\sigma_0 \phi t_c} \right\}^{\frac{1}{2}} \exp(\phi/2k_0 T). \qquad (6.48)$$

It is clear that simple forms of the relation (6.45), other than that assumed in (6.46), merely alter the result (6.48) by a factor of order of magnitude unity.

Since the conductivity law (1.2) and the inequalities used in calculating the approximate expression (6.48) are realistic for many dielectrics, it is of interest to investigate why the critical field strength is independent of the choice of T_m. To this end we find an approximate expression for the rate of rise of temperature. Putting (6.46) into (6.44) we have

$$C_V \frac{dT}{dt} = \sigma_0 \left(\frac{F_c}{t_c} \right)^2 t^2 \exp(-\phi/k_0 T). \qquad (6.49)$$

Separation of the variables and integration yields

$$(k_0 T_0^2/\phi)\{\exp(\phi/k_0 T_0) - \exp(\phi/k_0 T)\} \simeq (\sigma_0 F_c^2/3C_V t_c^2)t^3, \qquad (6.50)$$

the validity of this equation relying on the condition $\phi \gg k_0 T$ for all temperatures. If T is of order T_0, a simple solution of eqn (6.50) gives

$$T = T_0 \left\{ 1 + \frac{\sigma_0}{3C_V T_0} \left(\frac{F_c}{t_c} \right)^2 t^3 \exp(-\phi/k_0 T_0) \right\}. \tag{6.51}$$

Putting $t = t_c$ and using the result (6.48) for F_c in (6.51), we have for the approximate expression for the temperature attained over the critical time

$$T \simeq T_0(1 + k_0 T_0/\phi), \tag{6.52}$$

showing a very small rise over the ambient.

The impulse thermal condition then requires that the temperature given by (6.52) be exceeded, and not that T_m be exceeded. However, both conditions will be satisfied approximately, since almost all of the pulse time is occupied in raising the temperature to the value given by (6.52), and the further, very large temperature rise occurs in a very short time.

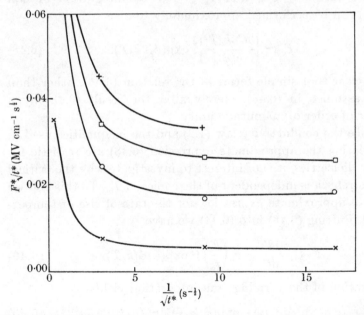

FIG. 6.8. Plot of breakdown strength as a function of voltage rise time in the form $F^*\sqrt{t^*}$ against $1/\sqrt{t^*}$ for NaCl at 350 °C and 200 μm thickness. + Nominally pure (Source 1). □ Nominally pure (Source 2). ○ $1 \cdot 25 \times 10^{-5}$ mole fraction Mn^{++}. × 8×10^{-4} mole fraction Mn^{++}. (After Hanscomb 1962.)

Evidence of impulse thermal breakdown in sodium chloride was found by Andreev (1958), Hanscomb (1962), and Heyes and Watson (1968). The case is most clearly presented by Hanscomb, who investigated the breakdown of sodium chloride at 350 °C using voltage wave forms rising linearly with time. His results, shown in Fig. 6.8, give the breakdown strength as a function of voltage-rise time in a form which allows for easy comparison with the impulse thermal critical field strength (6.48). Over a range of breakdown times from 3·5 to 100 ms, $F^*\sqrt{t^*}$ is approximately constant for each state of crystal purity. After measuring the electrical conductivity of the crystals concerned, Hanscomb (1962) found that breakdown strengths differed from the critical field strengths of eqn (6.48) by less than 5 per cent for measurements at 3·5 and 12 ms. This appears to establish the impulse thermal mechanism as being the operative one under these circumstances.

A further interesting fact has been demonstrated by Hanscomb (1962), who showed that for all specimens and rise times in his work the breakdown strength was proportional to the inverse square root of the conductivity at ambient temperature; this is illustrated in Fig. 6.9. Whitehead (1951) gives this

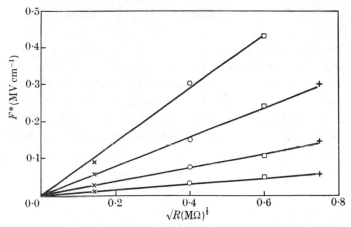

FIG. 6.9. Breakdown strength of NaCl as a function of the square root of specimen resistance at ambient temperature, for various voltage rise times and purities (the purity is indicated in the same way as in Fig. 68). Specimen thickness was 200 μm and temperature 350 °C. (After Hanscomb 1962.)

relation between breakdown strength and ambient conductivity as a general law of thermal breakdown—reference to the steady-state and impulse-thermal solutions (eqns (6.14), (6.31), and (6.48)) shows that it is obeyed in all the limiting cases discussed above, and it seems reasonable to presume that it will hold in intermediate cases. There is thus good evidence to conclude that all of the breakdown observed by Hanscomb (1962) was thermal in nature. Konorova and Sorokina (1965) argue that the proportionality between breakdown strength and inverse square root of the conductivity extends down to 50 °C for various alkali halides, provided that the field dependence of the conductivity is taken into account.

Hanscomb (1969) also investigated the thermal breakdown of NaCl and KCl at temperatures between 200 °C and 320 °C, using flat-topped pulses of voltage. In comparison with theoretical results, account was taken of the field dependence of the conductivity by using the expression (6.42). The relationship between critical field strength and time to breakdown is easily derived from eqn (6.44) and is

$$t_c \simeq C_V k_0 T_0^2 / \phi \sigma(T_0, F_c) F_c^2. \tag{6.53}$$

Hanscomb's data for nominally pure KCl specimens are shown in Fig. 6.10 together with theoretical results based on eqn (6.53). Similar results were also found for KCl doped with $SrCl_2$, and also for NaCl both pure and $MnCl_2$ doped. It is noteworthy that the experimental results diverge sharply from the theoretical curves when $t^* > 10$ ms; this effect is presumed to be due to the neglect of the heat-conduction term in deriving (6.53). It is easily seen from the general equation of thermal breakdown (6.1) that the impulse-thermal assumption can be expected to be valid only subject to the inequality

$$t_c \lessapprox C_V l^2 / \kappa, \tag{6.54}$$

where l is a length of order of the thickness of the dielectric specimen if the electrodes are good thermal sinks, but greater otherwise. With $C_V \sim 1$ J cm^{-3}, $l \sim 10^{-2}$ cm, and $\kappa \sim 10^{-1}$ J

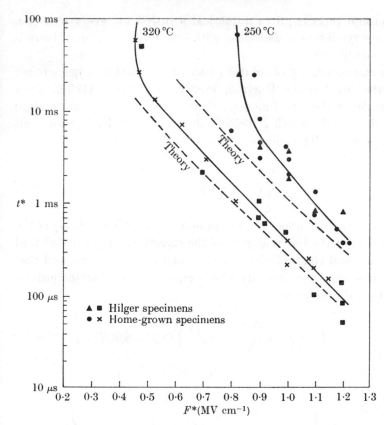

FIG. 6.10. Variation of the time to breakdown with breakdown strength for pure KCl at 250 °C and 320 °C with single flat-topped voltage pulses. Calculations based on eqn (6.54) are shown for comparison with the experimental results. (After Hanscomb 1969.)

cm^{-1} s^{-1} °C^{-1}, eqn (6.54) yields $t_c \lesssim 10^{-3}$ s, which is satisfactory order-of-magnitude agreement with Hanscomb's results.

6.4. Thermal breakdown for arbitrary times

The types of thermal breakdown so far considered correspond to the limiting cases of infinite and infinitesimal time of voltage application. For each of these limiting cases, a term drops out of eqn (6.1), and the resulting theory can be handled analytically up to a point. If no such simplifying assumptions are possible, then eqn (6.1) requires numerical solution for every case (i.e. for every different dielectric with every different set of boundary conditions).

Some generality of solution may be retained in a special case treated by Copple, Hartree, Porter, and Tyson (1939). They considered the one-dimensional problem (i.e. the infinite slab) for the case in which the electrodes are maintained at ambient temperature. The equation (6.1) then becomes

$$C_V \frac{\partial T}{\partial t} - \frac{\partial}{\partial z}\left(\kappa \frac{\partial T'}{\partial z}\right) = \sigma \left(\frac{V}{d}\right)^2, \qquad (6.55)$$

for which, by virtue of the boundary conditions, (6.14) is the solution in the limiting case of the steady state, provided that the thermal conductivity is temperature-independent and that the electrical conductivity obeys eqn (1.2). This latter proviso further simplifies (6.55) to

$$C_V \frac{\partial T}{\partial t} - \kappa \frac{\partial^2 T}{\partial z^2} = \sigma_0 \left(\frac{V}{d}\right)^2 \exp(-\phi/k_0 T), \qquad (6.56)$$

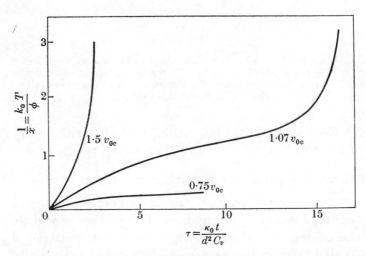

FIG. 6.11. Diagram showing the numerical solution of eqn (6.57) for various values of applied voltage (after Copple et al. 1939).

which can be written in a dimensionless form,

$$\frac{\partial(1/x)}{\partial\tau} - \frac{\partial^2(1/x)}{\partial\zeta^2} = 8\left(\frac{V}{V_{oc}}\right)^2 \frac{1}{x_0^2} \exp(x_0 - x), \qquad (6.57)$$

where we have used eqns (6.14) and (6.23) and also introduced the dimensionless quantities

$$\tau = \kappa t / d^2 C_V \qquad (6.58)$$

and

$$\zeta = z/d. \qquad (6.59)$$

The numerical solution of eqn (6.57) gave the results shown in Fig. 6.11, whose general features have been foreshadowed in Fig. 1.4.

PURELY ELECTRICAL BREAKDOWN

7.1. Intrinsic critical field strengths

IT has been pointed out in §1.3 that the experimental concept of purely electrical breakdown is that of a breakdown phenomenon which is not connected in any way with the Joule heating effect of the pre-breakdown current. The instability, which undeniably has thermal effects, develops very rapidly but is not of thermal origin.

Just as in the case of high-field conduction, there have been bulk-limited theories, electrode-limited theories, and combinations of the two. The usual calculations of intrinsic critical field strengths are in the former category, while avalanche and space-charge theories are of the latter type.

(a) *Theories based on the single-electron approximation*

The simplest theories of intrinsic critical field strength consider the average behaviour of a single conduction electron in a strong electric field. For this consideration to be valid, it is required that the density of conduction electrons be so low that only the applied field and the interaction with the lattice determine the motion of the electron. The energy-balance equation (1.9) (on which any calculation of critical intrinsic field strength must be based) then becomes

$$A(F, T, E) = B(T, E) \qquad (7.1)$$

where the parameter α of (1.9) becomes E, the energy of the electron whose average behaviour is being considered. A difficulty immediately arises concerning each of the terms of (7.1), viz. that the average rates of energy transfer represented by A (the gain from the applied field) and B (the loss to the lattice) are capable of a meaningful interpretation only if the energy change per inelastic collision is much smaller than the electron energy. Thus

$$E \gg \hbar\omega \tag{7.2}$$

is a strict requirement, and the critical-field criterion must be framed so as to consider only those electrons for which (7.2) holds.

From eqn (2.43) the rate of energy gain per unit volume from the applied field is

$$A(F, T, E) = e^2 F^2 \tau(E, T)/m^*, \tag{7.3}$$

and $\tau(E, T)$ is given by either (2.55) or (2.61) for a polar or a non-polar crystal.

The average rate of energy loss to the lattice can be calculated from

$$B(E, T) = \sum_{\mathbf{w}} \hbar\omega(\mathbf{w})(P_{\mathbf{w}}^e - P_{\mathbf{w}}^a), \tag{7.4}$$

where the lattice frequency $\omega(\mathbf{w})$ is a function of the wave vector $\mathbf{w}$, and the emission and absorption probabilities are defined by eqn (2.35). Using (2.37) in (7.4) and replacing the sum by an integral, we get

$$B(E, T) = \frac{Vm}{2\pi k\hbar^3}\Big\{ \int wG(w)\hbar\omega(n_w+1) \, dw -$$
$$- \int wG(w)\hbar\omega n_w \, dw \Big\}. \tag{7.5}$$

If the discussion is restricted to polar crystals (which have been the subject of almost all applications of intrinsic critical-field calculations) the form (2.48) can be used for the interaction constant; in conjunction with (2.56) and (2.57) we obtain from (7.5)

$$B(E, T) = \frac{\hbar\omega}{2\tau_0(E)} \int \frac{dw}{w}. \tag{7.6}$$

The lower limit of integration cannot be put at zero as in (2.52), since the most important contributions come from small wave numbers. The conservation of energy

$$(\hbar^2 k^2 - \hbar^2 k'^2)/2m = \hbar\omega$$

can be used in the limiting cases

$$k+k' \simeq 2k$$

and

$$k-k' \simeq w_{min}$$

to give the lower limit in (7.6). The result is

$$w_{min} = m\omega/\hbar k, \tag{7.7}$$

which when used in conjunction with the upper limits for w used in (2.52) gives

$$B(E, T) = \frac{\hbar\omega}{2\tau_0(E)} \ln \gamma, \tag{7.8}$$

where

$$\gamma = \pi(2E)^{\frac{1}{2}}/a\omega m^{\frac{1}{2}} \tag{7.9}$$

and $\tau_0(E)$ is given by (2.56) for electron energies greater than the value E_0 of (2.54). For lower electron energies

$$\gamma = 4E/\hbar\omega, \tag{7.10}$$

and $\tau_0(E)$ is given by (2.57).

This simply derived result for the rate of energy loss has received qualitative confirmation by a numerical calculation of Thornber and Feynman (1970) for the Fröhlich polaron model. In both cases the energy loss is a maximum for electrons having energy of order of the lattice quantum.

Although we have examined a model of an alkali halide having a single relaxation frequency ω for longitudinal polarization waves, it is not unreasonable to use different values of ω in eqns (2.55) and (7.8). This situation arises since the short-wavelength phonons contribute most to the relaxation time, so that we should put

$$\omega = \omega_t \tag{7.11}$$

in (2.55), where ω_t is the Reststrahlen angular frequency. On the other hand, energy loss to the lattice is determined mainly by interaction with the long-wavelength phonons, for which we should put

$$\omega = \omega_t(\epsilon_s/\epsilon_\infty)^{\frac{1}{2}}. \tag{7.12}$$

The development of a critical field-strength criterion from the energy-balance equation (7.1) amounts to an identification of an energy E for which this energy balance is to hold, and then calculation of F_c from the resulting equation. The high-energy criterion was formulated by Fröhlich (1937) and is based on the idea that indefinitely large multiplication of electrons in the conduction levels will destroy the dielectric. The quantities $A(F, T, E)$ of (7.3) and $B(E, T)$ of (7.8) are shown schematically in Fig. 7.1. The high-energy critical field-strength criterion

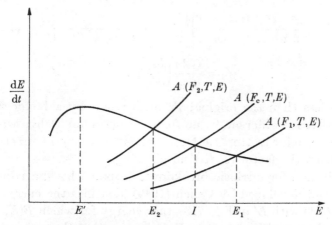

Fig. 7.1. The average rate of energy gain $A(F, T, E)$ from an applied field for various field strengths, and the average rate of energy loss to the lattice $B(E, T)$.

may be stated as

$$A(F_c, T, I) = B(I, T), \qquad (7.13)$$

where I is the ionization energy corresponding to a transition from the valence band to the conduction band. In order to see that this equation does in fact represent a critical situation, consider three possible values of the field strength, F_1, F_2, and F_c, such that $F_1 < F_c < F_2$ as shown in Fig. 7.1. In an applied field F_1 an electron must have an energy E_1 if it is to be, on the average, accelerated by the field; for the field F_2 the critical energy is E_2. The equation (7.13) then determines the critical field strength as being that for which collision ionization from the valence band cannot be balanced by the inverse process of recombination. That F_c is such a critical field strength is evident from the fact that the high-energy electron resulting from a recombination collision is not able to lose its energy to the lattice if the applied field exceeds F_c. Since it is, on the average, accelerated by the field, it will cause a further ionizing collision and thus nullify the effect of the recombining collision as a means of removal of electrons from the conduction band. The critical field strength can then be determined by using eqns (2.55), (2.56), (7.3), (7.8), and (7.9) in (7.13); the result is

$$F_c = \frac{\pi^2}{2} \frac{e^3}{a^4 I M} \left\{ \frac{m\hbar \ln \gamma'}{\omega_t} \right\}^{\frac{1}{2}} \left\{ 1 + \frac{2}{\exp(\hbar\omega_t/k_0 T) - 1} \right\}^{\frac{1}{2}}, \quad (7.14)$$

where

$$\gamma' = \frac{\sqrt{2}\pi}{a\omega_t} \left(\frac{I\epsilon_\infty}{m\epsilon_s} \right)^{\frac{1}{2}}. \quad (7.15)$$

This is less than the original Fröhlich result by a factor $2^{-\frac{1}{3}}$ arising from a different value for the maximum Debye wave number w_0 (cf. Stratton 1961); a further very small correction arises from the factor $(\epsilon_\infty/\epsilon_s)^{\frac{1}{2}}$ in γ'.

A criterion for intrinsic breakdown proposed by von Hippel (1935) and developed by Callen (1949) identifies the energy E of eqn (7.1) with E' of Fig. 7.1—that energy for which $B(E, T)$ is a maximum with respect to E. It is clear that the evaluation of F_c from

$$A(F_c, T, E') = B(E', T) \quad (7.16)$$

amounts to finding a critical field that is able, on the average, to accelerate all conduction electrons against the retarding influence of the lattice. Clearly, eqn (7.16) results in a more stringent criterion for breakdown than does (7.13). The use of eqns (2.55), (2.57), (7.3), (7.8), and (7.10) in (7.16) yields for the critical field strength

$$F_c = \frac{em}{2\sqrt{E'}} \frac{1}{\epsilon^*} \left(\frac{\omega_t^3 \epsilon_s^3 \ln \gamma''}{\hbar\epsilon_\infty^3} \right)^{\frac{1}{2}} \times$$

$$\times \left\{ 1 + \frac{2}{\exp(\hbar\omega_t \epsilon_s^{\frac{1}{2}}/k_0 T \epsilon_\infty^{\frac{1}{2}}) - 1} \right\}^{\frac{1}{2}} \quad (7.17)$$

where

$$\gamma'' = \frac{4E'}{\hbar\omega_t} \left(\frac{\epsilon_\infty}{\epsilon_s} \right)^{\frac{1}{2}}. \quad (7.18)$$

This formula is in a more simple form than Callen's original result, but the two are numerically approximately equal in most cases.

(b) Collective critical-field theories

If the electron distribution function is taken into account, the calculation of the critical field becomes more difficult in principle. However, the assumption of a Maxwellian distribution of conduction electrons possessing a temperature T_e greater than the lattice temperature T enables one to write the energy-balance equation (1.9) as

$$A(F, T, T_e) = B(T, T_e), \qquad (7.19)$$

the parameter α being in this case the conduction-electron temperature.

The electronic-conduction properties of dielectrics having a Maxwell distribution of conduction electrons were studied by Fröhlich and Paranjape (1956) for the case of a trap-free crystalline dielectric (cf. §4.2(b), (c)) and by Fröhlich (1947a) for an amorphous dielectric containing traps (cf. §4.3(a)). In both cases, equations were derived ((4.38) and (4.61)) giving the electron temperature as a function of the applied field; in both of these cases also a stable solution was possible only up to a certain critical electronic temperature T_c, which defines a critical value F_c of the applied field.

Fröhlich and Paranjape (1956) and Stratton (1961) have shown that the critical field strength derived from (4.38) for ionic crystals is given by

$$F_c = G(T/\Theta)F_0, \qquad (7.20)$$

where

$$F_0 = \frac{m e \omega_t}{\hbar}\left(\frac{\epsilon_s}{\epsilon_\infty}\right)^{\frac{1}{2}} \frac{1}{\epsilon^*} \qquad (7.21)$$

is a field strength characteristic of the dielectric and G is a function of T/Θ alone. The temperature Θ is defined by

$$\Theta = \frac{\hbar \omega_l}{k}\left(\frac{\epsilon_s}{\epsilon_\infty}\right)^{\frac{1}{2}}. \qquad (7.22)$$

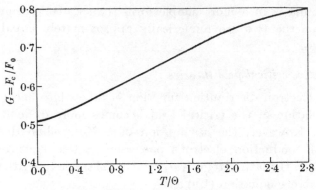

FIG. 7.2. The function $G(T/\Theta)$ of eqn (7.20) (after Stratton 1961).

The function G is of order unity and increases slowly with increasing temperature; it is shown in Fig. 7.2.

The application of the theory to non-polar crystals is most conveniently expressed in terms of the mobility at low field strengths. The results are given by Stratton (1961):

$$F_c \simeq 0 \cdot 30 \frac{\hbar N^{\frac{1}{3}}}{m\mu(\Theta)}\left(\frac{T}{\Theta}\right)^{\frac{1}{2}} \quad \text{if} \quad T \gg \Theta, \qquad (7.23)$$

and

$$F_c \simeq 0 \cdot 14 \frac{\hbar N^{\frac{1}{3}}}{m\mu(\Theta)} \quad \text{if} \quad T \ll \Theta, \qquad (7.24)$$

in which N is the number of atoms per unit volume, and the Debye temperature is defined for a non-polar crystal by

$$\Theta = (6\pi^2 N)^{\frac{1}{3}}\hbar s/k. \qquad (7.25)$$

For the Fröhlich amorphous-dielectric model, the appropriate critical field strength has already been derived algebraically in eqn (4.65). This theory of intrinsic breakdown gives a temperature dependence rather than an absolute value for the critical field strength; the other theories all yield absolute values without disposable constants.

(c) Intrinsic critical field strengths for alkali halides

The critical field strengths for the Fröhlich high-energy criterion (eqn (7.14)), the von Hippel–Callen low-energy

criterion (eqn (7.17)), and the Fröhlich–Paranjape collective
criterion (eqn (7.20)) are plotted as functions of temperature
for NaCl, KCl, and KBr in Fig. 7.3. The values of the relevant
crystal constants are listed in Appendix 1. The numerical
results of Fig. 7.3 differ slightly from those quoted in original
papers, either because of different values of crystal constants or
for reasons given in the text. The breakdown field strengths
measured by various workers are also shown in Fig. 7.3.
Results due to Cooper and co-workers are indicated by vertical

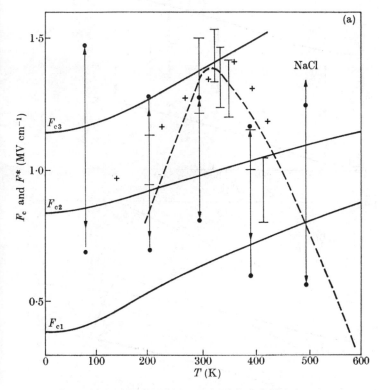

FIG. 7.3. Breakdown field strengths and critical field strengths over a range
of temperature for (a) NaCl, (b) KCl, (c) KBr. ○ Austen and Whitehead
1940, – – Alger and von Hippel 1949, ⊤ Calderwood and Cooper 1953,
⁞ Cooper et al. 1957, ↕ Cooper et al. 1960, × Konorova and Sorokina 1957,
+ Kuchin 1957. Critical fields are F_{c1}, Fröhlich high-energy criterion, F_{c2},
von Hippel–Callen low-energy criterion, F_{c3}, Fröhlich–Paranjape hot-electron
theory.

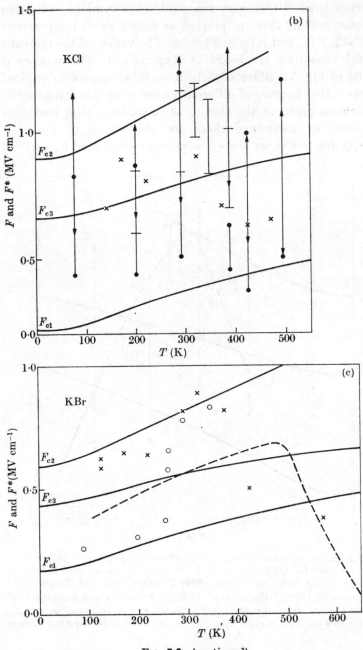

FIG. 7.3. (continued)

lines showing the range of variation of the observations, while the points attributed to other workers are average values of a number of measurements. Where breakdown strengths were measured for various rise times of the applied voltage, the values for d.c. breakdown (or for the longest rise time) are given in Fig. 7.3. In addition, the results of Cooper, Grossart, and Wallace (1957) are those obtained for breakdown along the (100) crystal direction, while those of Cooper, Higgin, and Smith (1960) apply to the (110) direction. All of the results quoted are for crystals which were pure, presumably to laboratory reagent standard or better.

It requires no detailed study of Fig. 7.3 to see that we do not have an experimental situation that is adequately explained by a theory. One reason for this is the lack of reproducibility of the experimental results, but even if reproducibility of results is attained it is unlikely that any of the theoretical critical fields will adequately explain them, since these critical fields refer to a homogeneous infinite medium without consideration of the boundary conditions. It is true, however, that the experimental results almost all lie within the boundaries of the critical fields quoted, and since the latter are calculated without any disposable constants it is highly likely that the calculations are in some way relevant. Since the common thread in the calculation of intrinsic critical fields is the onset of significant collision ionization when those critical fields are exceeded, it seems that the conclusion to be drawn from the data and theories exhibited in Fig. 7.3 is that significant collision ionization is an important factor in the onset of purely electrical breakdown.

7.2. Avalanche critical field strengths

In theories of the intrinsic critical field an increase in the number of conduction electrons is viewed as a sudden occurrence following a situation in which a steady state obtained. On the other hand, avalanche critical-field theories consider that this conduction-electron multiplication is a process that gradually reaches intolerable proportions as the field strength rises. The meaning of the phrase 'intolerable proportions' depends on the criterion adopted, e.g. the impulse thermal criterion of §6.3 may be applied to the avalanche current, or some purely arbitrary condition may be assigned. It is pointed out in §1.3 that the two basic theoretical assumptions used to simplify calculations are uniformity of field and continuity of current. In this section, we shall confine our attention to theories based on a uniform field, and defer till the following section considerations of space-charge distortion of the field.

Avalanche critical-field theories can be divided on the score of the mechanism of initiation of the avalanche. The first type that we consider is that in which the avalanche of electrons arises by field emission from the valence to the conduction band; this has been called field-emission breakdown and was first considered by Zener (1934). The second manner of initiation of the avalanche is by collision ionization, and there have been various critical criteria formulated on the basis of this mechanism.

(a) Field-emission critical field strengths

The field-emission critical field strength introduced by Zener (1934) was based on the calculation of the probability per unit time that an electron would escape from the valence to the conduction band. He obtained the Fowler–Nordheim-type result

$$P_{vc} = \frac{eFa}{h} \exp\left\{-\frac{\pi^2 maI^2}{h^2 eF}\right\}, \tag{7.26}$$

where I is the ionization energy. Substitution of reasonable values for the quantities in (7.26) shows that the exponent does not reduce to unity until field strengths of the order of tens of megavolts per centimetre have been attained. It is then argued that the ionization probability is very small unless such a condition applies, so that eqn (7.26) gives a basis for an order-of-magnitude estimate of a critical field strength.

Franz (1939) has put this idea on a more quantitative basis by using the impulse thermal-breakdown theory as the test of criticality of the field-emitted avalanche. If the effect of

Coulomb interaction between the emitted electron and the valence-band hole is taken into account, the transition probability becomes

$$P_{\mathrm{vc}} = \gamma F^{10/3} \exp(-\beta/F) \qquad (7.27)$$

(cf. Franz 1956), where

$$\beta = \pi(2m^*)^{\frac{1}{2}} I^{\frac{3}{2}}/4\hbar e \simeq 4 \times 10^7 (m^*/m)^{\frac{1}{2}} I^{\frac{3}{2}}, \qquad (7.28)$$

in which m^* is the effective mass at the top of the valence band or the bottom of the conduction band (assumed equal) and the energy gap I is in electronvolts. The constant γ is given approximately by $\gamma \simeq 10^{-7}$ s^{-1} V$^{-10/3}$ cm$^{10/3}$. Consider now the situation in which a field F_{c} is applied for a time t_{c}, the combination being critical in the impulse thermal sense, so that

$$C_V(T_m - T_0) = \int_0^{t_{\mathrm{c}}} jF_{\mathrm{c}}\, \mathrm{d}t = e\mu F_{\mathrm{c}}^2 \int_0^{t_{\mathrm{c}}} n(t)\, \mathrm{d}t, \qquad (7.29)$$

where $n(t)$ is the density of conduction electrons at time t. If it is assumed that the process of breakdown is complete before an appreciable number of field-emitted electrons have entered the anode, then

$$\mathrm{d}n(t)/\mathrm{d}t = P_{\mathrm{vc}} N_{\mathrm{v}}, \qquad (7.30)$$

where N_{v} is the valence-electron density. The condition of validity of (7.30) is

$$t_{\mathrm{c}} \ll d/\mu F_{\mathrm{c}}, \qquad (7.31)$$

where d is the inter-electrode distance. Substituting (7.27) and (7.30) in (7.29), we have

$$C_V(T_m - T_0) = \tfrac{1}{2}\gamma N_{\mathrm{v}} e\mu F_{\mathrm{c}}^{16/3} t_{\mathrm{c}}^2 \exp(-\beta/F_{\mathrm{c}}). \qquad (7.32)$$

This equation can be solved by successive approximations in any given case, but a first approximation is easily given since the only parameter sensitively affecting the result is β. Equation (7.32) can be rewritten

$$F_{\mathrm{c}} = \beta/\ln\left\{\frac{e\alpha\mu F_{\mathrm{c}}^{16/3} N_{\mathrm{v}} t_{\mathrm{c}}^2}{2C_V(T_m - T_0)}\right\}. \qquad (7.33)$$

Taking as a trial solution of eqn (7.33) $F_{\mathrm{c}} = 10^7$ V cm^{-1} and using $C_V(T_m - T_0) \sim 10^{22}$ eV cm^{-3}, $N_{\mathrm{v}} \sim 10^{23}$ cm^{-3}, $\mu \sim 10$ cm^2

$V^{-1} s^{-1}$, $m^* \sim m$, we have, using (7.28),

$$F_c \simeq \frac{4 \times 10^7 I^{\frac{3}{2}}}{\ln (10^{20} t_c^2)},\tag{7.34}$$

where I is in eV and t_c in μs. Equation (7.34) will give a critical field strength in substantial agreement with the trial value over a wide range of voltage application times in the μs region. Results calculated from (7.34) should be valid for several orders of magnitude above and below the trial solution, since the next step in the successive-approximation method will alter the denominator of (7.34) by a relatively small amount.

(b) The forty-generations critical field strength

The so-called forty-generations critical field strength is based on a single-electron collision-ionization theory originally developed by Fröhlich (1940) and by Seitz (1949) and formulated in a somewhat more general way by Stratton (1961), whose approach we shall follow.

Consider the simple picture of avalanche formation in which one electron, initially at the cathode, receives sufficient energy from the applied field to ionize a bound electron. If both of these electrons now receive the same energy from the field, they will each cause a further ionization; eventually 2^i free electrons will be formed if there are i generations between cathode and anode. The avalanche will lead to breakdown if i is sufficiently large, and to estimate this critical avalanche size we use the simple argument of Seitz (1949). Assume that the critical field strength is of order 10^6 V cm^{-1} and that the average mobility is of order 1 cm^2 V^{-1} s^{-1} at this field strength, which means that the electron travels about 1 cm in 1 μs. In addition to moving in the direction of the field, the electron diffuses in a plane normal to the field direction with a diffusion coefficient of order 1 cm^2 s^{-1}, so that the electron wanders in a cylinder of radius 10^{-3} cm during a 1 μs drift. We assume that all the electrons of the avalanche also lie within a cylinder of 10^{-3} cm radius while drifting 1 cm in the field direction. There are about 10^{17} atoms in a tube of a solid 1 cm long and 10^{-3} cm in radius, and energy of order of 10 eV per atom would disrupt the structure of the material. Assuming that none of the energy supplied by the field is conducted away from the tube thermally, the above data show that about 10^{18} eV would be required to cause disruption in the tube. With a field of 10^6 V cm^{-1}, this requires an avalanche of 10^{12} electrons, or about forty generations. It is clear that this analysis is extremely rough and approximate, but we accept it for the purpose of the simple theory since a much more elaborate analysis (cf. Stratton 1961) leads to a figure of thirty-eight generations for critical conditions. In any case, the resulting critical field strength is not at all sensitive to the exact value of the critical number of generations, and we therefore adopt the value $i = 40$ for criticality.

We turn now to the process by which an electron acquires sufficient energy to cause an ionizing collision. An expression for the collision ionization rate per unit time has already been introduced (eqn (5.27)); it is similar to, but not in exact agreement with, a result for the collision ionization rate per unit length

$$\alpha = \alpha_n \exp(-H/F),\qquad(7.35)$$

in which α_n is now a quantity having the dimension of an inverse length. Since the drift velocity of an electron in an applied field is given by $\langle\mu\rangle F$, eqns (5.27) and (7.35) are inconsistent; however, the rapidly varying exponential term is such that slight adjustments in the parameters α_n and H would suffice to make these two empirical equations into approximately equivalent descriptions of the collision ionization process.

A simple justification for the form (7.35) was given by Shockley (1961). Consider the probability $P(l)$ that a conduction electron has suffered no collisions with the lattice over a path length l; then the probability that it should suffer none in the subsequent path interval dl is $(1-dl/\lambda)$, where λ is the mean free path, which is in general a function of the energy. In view of the definition of $P(l)$, we have

$$P(l+dl) = P(l)(1-dl/\lambda)$$

or
$$dP/dl = -P(l)/\lambda. \tag{7.36}$$

The solution of eqn (7.36) is

$$P(l) = \exp\left(-\int_0^l dl/\lambda\right). \tag{7.37}$$

The collision free path length required to produce collision ionization is given by
$$l_I = I/eF. \tag{7.38}$$

If we assume that λ can be taken outside the integral in eqn (7.37) and replaced by some suitably averaged value $\langle\lambda\rangle$, then the use of (7.38) in (7.37) yields the probability that an electron makes no collision with the lattice over a path interval

$$P(l_I) = \exp(-I/eF\langle\lambda\rangle). \tag{7.39}$$

If one assumes that all of the field dependence of the ionization is contained in (7.39), comparison with (7.35) yields

$$H = I/e\langle\lambda\rangle. \tag{7.40}$$

A more exact discussion of this problem was given by Baraff (1962), who solved the appropriate Boltzman equation numerically for the case of constant mean free path and quasi-free electrons. The results of his computations for the ionization function are shown in Fig. 7.4. It is clear from Baraff's results that the pre-exponential term depends on the field strength in a non-negligible way unless the optical phonon energy exceeds about 3 per cent of the ionization energy. This condition would almost never apply in an insulator, so that eqns (5.27) and (7.35) with H given by (7.40) must be regarded as rough approximations indeed. However, the approximately linear nature of Baraff's curves means that eqns (5.27) and (7.35) can be used if H is treated as an empirical constant that is not to be identified with the quantity given by eqn (7.40).

In the forty-generations theory, the critical ionization rate will be given by
$$\alpha_c d = 40 \tag{7.41}$$

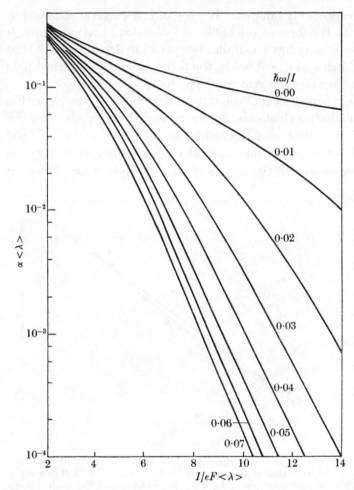

Fig. 7.4. Dependence of the ionization α-function on the applied field F for various values of the optical phonon energy $\hbar\omega$. The average value of the mean free path is denoted by $\langle\lambda\rangle$ and the ionization energy by I. (After Baraff 1962.)

where d is the inter-electrode distance. With the use of eqn (7.35), eqn (7.41) yields for the critical field strength

$$H/F_0 = \ln(\alpha_n d/40). \qquad (7.42)$$

This is a less elaborate equation for the forty-generations theory than has usually been presented (cf. Stratton 1961), but it is probably adequate in view of the uncertainties associated with the form of the collision ionization function. Since H will presumably have a similar temperature dependence to that of the intrinsic critical fields, the forty-generations critical field will be likewise temperature-dependent.

Since the reciprocal of the critical field varies as the logarithm of the dielectric thickness, comparison with experimental work necessitates that measurements be available over at least several decades of thickness. Such extensive data is rather rare, but compilations of the results of various workers are shown for

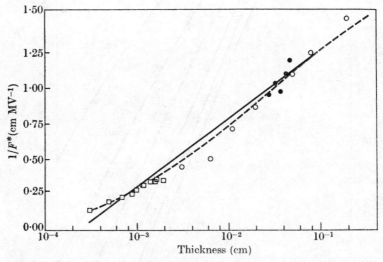

FIG. 7.5. The reciprocal of the mean breakdown field strength of NaCl as a function of the logarithm of the inter-electrode distance. The full line represents a fit to the forty-generations theory (eqn (7.42)) and the dashed line a fit to the space-charge-enhanced cathode-emission theory of Fig. 7.8. Experimental points are from ○ Watson, Heyes, Kao, and Calderwood 1965, ● Cooper and Smith 1961, □ Vorob'ev, Vorob'ev, and Murashko 1963.

NaCl in Fig. 7.5 and for Al_2O_3 in Fig. 7.6. In both cases, the reciprocal of the breakdown field strength shows an approximate linear variation with the logarithm of the thickness. Fitting of the data to eqn (7.42) yields the values for the ionization parameters shown in Table 7.1. The figures are not entirely satisfactory, since the ionization rate for Al_2O_3 at the highest

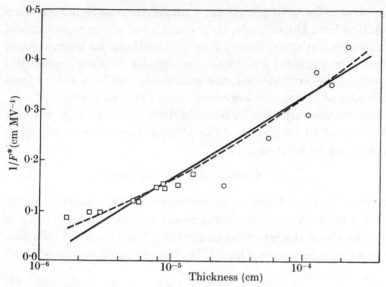

FIG. 7.6. The reciprocal of the mean breakdown field strength of Al_2O_3 as a function of the logarithm of the inter-electrode distance. The full line and the dashed line have the same significance as in Fig. 7.5. Experimental points are from ○ Merrill and West 1963, □ Lomer 1950.

field strength for which data are given in Fig. 7.6 corresponds to a good deal more expenditure of energy than is available from the applied field.

(c) Avalanche-enhanced cathode emission

Critical conditions of another type were considered by Forlani and Minnaja (1964). Rather than taking account of

TABLE 7.1

Values of ionization parameters obtained from a fit of the forty-generations theory to experimental results.

Ionization parameters	Substance	
	NaCl	Al_2O_3
α_n cm^{-1}	$1 \cdot 0 \times 10^5$	4×10^7
H MV cm^{-1}	$5 \cdot 0$	14

the avalanche multiplication resulting from only one electron starting from the cathode, they considered the consequences of avalanche multiplication of Fowler–Nordheim emission current from the cathode. Although the details of their theoretical approach are complicated, the final result can be obtained from an order-of-magnitude argument (cf. O'Dwyer 1969b).

The current injection into conduction levels from the cathode is assumed to be given by the Fowler–Nordheim result (3.1), which can be written

$$j_{\text{cath}} = j_0 \exp\{-4\sqrt{(2m^*)}\phi^{\frac{3}{2}}/3\hbar e F\}, \qquad (7.43)$$

where j_0 is a field-dependent pre-exponential quantity. If the lattice vibrations in the dielectric are completely ineffective in slowing down the electrons in a strong field, then the collision-ionization rate per unit length is given approximately by

$$\alpha = eF/I. \qquad (7.44)$$

The current arriving at the anode as a result of this collision-ionization multiplication will be given by

$$j_{\text{an}} = j_{\text{cath}}\exp(eFd/I) = j_0 \exp\left\{-\frac{4\sqrt{(2m^*)}\phi^{\frac{3}{2}}}{3\hbar e F} + \frac{eFd}{I}\right\}. \qquad (7.45)$$

If the somewhat arbitrary criterion is now adopted that zero exponent in eqn (7.45) corresponds to critical conditions, then the critical field strength is given by

$$F_{\text{c}}^2 = \frac{4\sqrt{(2m^*)}\phi^{\frac{3}{2}}I}{3\hbar e^2 d} . \qquad (7.46)$$

Inserting values of reasonable order of magnitude in (7.46), one readily finds

$$F_{\text{c}} \simeq 3 \times 10^4/d^{\frac{1}{2}} \text{ V cm}^{-1} \qquad (7.47)$$

with d measured in cm.

Experimental results consistent with eqn (7.46) have been found for various inorganic dielectrics by Budenstein and Hayes (1967), Smith and Budenstein (1969), and Budenstein, Hayes, Smith, and Smith (1969). A sample of their results for breakdown strength as a function of thickness is shown in

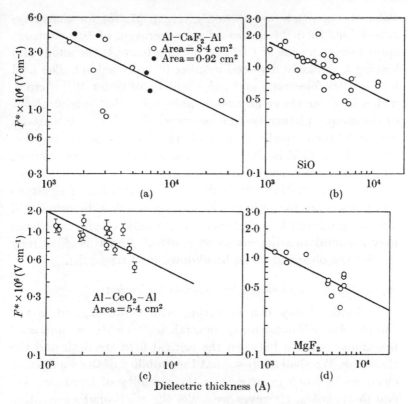

FIG. 7.7. Breakdown strength as a function of thickness for (a) CaF$_2$, (b) SiO, (c) CeO$_2$, (d) MgF$_2$. All data were taken at room temperature except for CeO$_2$ which was taken at 78 °K. The straight lines are drawn with a slope to agree with (7.46). (After Budenstein *et al.* 1969.)

Fig. 7.7. In order of magnitude the results agree with eqn (7.47), but it does seem that data over several decades of thickness would be required to establish a $d^{-\frac{1}{2}}$ dependence. For NaCl data is available over an extensive range of thickness (cf. Fig. 7.5) but in that case it does not agree with eqn (7.46).

7.3. Space-charge-enhanced critical fields

A theory that considers collision ionization as an important element and at the same time retains the assumption of a uniform field encounters severe difficulties if there are many generations of collision ionization (cf. O'Dwyer 1967). The order-of-magnitude calculation given in §7.2(b) for the forty-generations theory requires a final generation of 10^{12} electrons spread over an area of 10^{-5} cm² moving towards the anode and leaving a similarly disposed number of holes behind. The field between the electrons and holes will be of order 10^{11} V cm^{-1}, which is so far above any reasonable value that modifications of the simple picture cannot be expected to improve matters. The only theoretically simple alternative to the assumption of a uniform field is the assumption of continuity of current, and critical fields based on this assumption are discussed first.

Collision ionization with subsequent separation of electrons and holes is not the only mechanism by which a distorted field may be produced in a dielectric; in an ionic conductor charge may accumulate adjacent to an electrode and this effect may modify the observed mean breakdown field strengths.

(a) Space charge produced by collision ionization

A simple theory can be given, which, although it is not completely self-consistent, nevertheless yields a universal functional relation between the critical field strength and the thickness. We shall suppose that the mobility of the conduction electrons is much greater than the mobility of the holes, i.e. eqn (5.34) holds. However we make the additional assumption that the current is carried by electrons alone, which necessitates the more stringent condition

$$n\mu_n/p\mu_p \gg 1. \tag{7.48}$$

This inequality cannot remain true for extended periods of constant electron current, since the electrons continually create new holes by collision ionization; however it may be satisfied approximately over a short time interval. The continuity equation for electron current is then

$$ne\mu_n F = n\{1+\alpha(F)\,\mathrm{d}x\}e\mu_n(F+\mathrm{d}F), \tag{7.49}$$

where $\alpha(F)$ is the collision-ionization function (7.35). Expanding eqn (7.49) to first order we have

$$\mathrm{d}x = -\mathrm{d}F/F\alpha(F) \tag{7.50}$$

as the requirement for continuity of electron current alone. The field strength must of course be related to the carrier densities by Poisson's equation (5.2), but rather than bring it explicitly into the calculation we assume that the relatively immobile holes are arranged to satisfy it.

With eqn (7.35) for the collision-ionization rate per unit length, eqn (7.50) becomes

$$\mathrm{d}x = -\frac{\exp(-H/F)}{\alpha_n F}\,\mathrm{d}F \qquad (7.51)$$

Introducing a dimensionless distance†

$$X = \alpha_n x \qquad (7.52)$$

and the dimensionless field strength of (5.35), we can integrate eqn (7.51) to give

$$X = -\int_{\mathscr{F}_{\text{cath}}}^{\mathscr{F}} \frac{\exp(1/\mathscr{F})}{\mathscr{F}}\,\mathrm{d}\mathscr{F}. \qquad (7.53)$$

Equation (7.53) can be expressed formally in terms of the exponential integral $E_1(z)$ defined by eqn (5.38) to yield

$$X = E_1(1/\mathscr{F}) - E_1(1/\mathscr{F}_{\text{cath}}). \qquad (7.54)$$

For given values of the dielectric thickness, this equation predicts a very sharply rising value of $\mathscr{F}_{\text{cath}}$ when the mean field strength $\mathscr{F}$ passes a certain critical value, $\mathscr{F}_c$. This critical value of the dimensionless mean field strength can be computed as a function of dimensionless dielectric thickness (cf. O'Dwyer 1967) and the results are shown in Fig. 7.8. The fitting of this curve to experimental results for NaCl and Al_2O_3 is illustrated in Figs. 7.5 and 7.6; the collision-ionization parameters required to effect this fit are shown in Table 7.2. The values found are of the same order of magnitude as those calculated from a fit of the forty-generations theory; however, even for the highest field strengths considered, the data of

---

† This is not the same definition as (5.34), since α_n of (7.35) differs in meaning from α_n of (5.27).

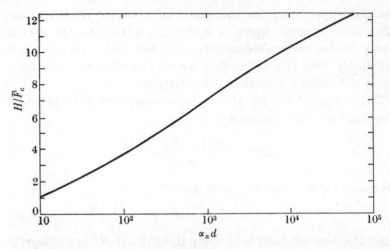

FIG. 7.8. Reciprocal of the dimensionless mean critical field strength as a function of dimensionless inter-electrode spacing for a simplified model of space-charge-enhanced cathode emission.

Table 7.2 do not yield impossibly high values of the ionization rate.

Vorob'ev and Pikalova (1969) have estimated the collision-ionization rate for NaCl. Over a limited range of field strength, they find a linear dependence on the field rather than the strongly varying dependence of (7.35). However their results agree to within about 30 per cent with those calculated from (7.35) and Table 7.2 within the range of field strength considered. The results mean that on the average between 100 eV

TABLE 7.2

Values of ionization parameters obtained from a fit of the theory of space-charge-enhanced cathode emission to experimental results

	Substance	
	NaCl	Al_2O_3
α_n (cm^{-1})	$2 \cdot 5 \times 10^4$	$7 \cdot 5 \times 10^6$
H (MV cm^{-1})	$6 \cdot 7$	20

and 200 eV of energy is acquired from the field for each collision-ionization event.

The collision-ionization rate for KBr was measured by Zelm (1968), using a technique in which electron avalanches were started by short light pulses. He found that the collision-ionization rate increased linearly from 20 to 50 cm^{-1} as the field strength increased from 1·5 to 4×10^5 V cm^{-1}.

Forlani and Minnaja (1969) have pointed out the difficulty in accepting the propostion that data on the thickness dependence of breakdown strength can be explained by the fitting of two parameters to the universal curve of Fig. 7.8. The difficulty with the simple theory lies in ignoring the contribution that the holes make to the total current. This defect can be corrected by using the theory of §§5.3 and 5.4(a); however, it is no longer possible to state the results in simple terms. Each individual situation must be calculated numerically.

The diagrams of Fig. 5.8 show a fairly sharply defined field strength that marks the onset of a current-controlled negative-resistance region. If this instability is identified with dielectric breakdown (a very attractive hypothesis in view of the demonstration by Ridley (1963) that high-current filaments form in regions of current-controlled differential negative resistance), we can plot the mean critical field strength as a function of thickness. This is done in Fig. 7.9 for the cases illustrated in Fig. 5.8(a) and (b). Also shown are the critical field strengths corresponding to hole mobility ten times greater than given by the data of (5.48); the higher hole mobility increases the critical field strength since it is more difficult to maintain a large space charge.

It is possible to make some progress analytically in breakdown theories governed by the analysis of §5.3, by exploiting the remarkable sameness of form of the curves shown in Fig. 5.5. This was done by O'Dwyer (1969c), who approximated the various computed results shown in Fig. 5.5 by the single family of curves shown in Fig. 7.10. In order to make progress with the problem, the emission characteristics of the cathode must be known, and also the fraction of total current that is electronic

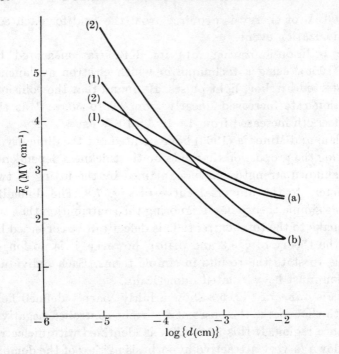

FIG. 7.9. Critical field strength as a function of inter-electrode spacing (1) for the case illustrated in Fig. 5.8 (a) and (b), and (2) for a hole mobility 10 times greater than that of case (1). Curves (a) refer to Fowler–Nordheim emission and curves (b) to Schottky emission.

at the cathode. The following simplifying assumptions are made.

(1) That the cathode emits with a Schottky-type characteristic

$$j = j_0 \exp[-\{\phi - (e^3 F/\epsilon)^{\frac{1}{2}}\}/k_0 T] \qquad (7.55)$$

in which the pre-exponential term is taken as approximately constant.

(2) That the current is totally electronic (see Fig. 5.4 for the validity of this assumption; it breaks down at all current levels for dimensionless thickness of 100 or greater).

Using the simplifications outlined above, we can find an approximate formula for the space-charge-enhanced cathode-emission critical field that will occur for sufficiently thick

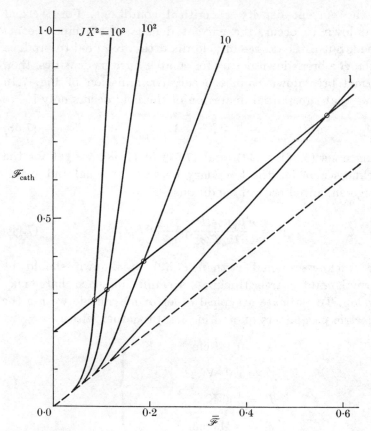

FIG. 7.10. Approximate synthesis of data of Fig. 5.5 for various values of the parameter JX^2 (cf. eqns (5.36) and (5.44)).

samples, and also a criterion to distinguish it from other types of breakdown that may occur when the samples are thin. Breakdown in the thin samples could, for example, be thermal, of the type discussed in §6.2(c). The energy-balance equation (6.26) leads with the use of (7.55) to

$$T_\mathrm{c} - T_0 = k_0 T_0^2 / \{\phi - (e^3 F/\epsilon)^{\frac{1}{2}}\} \qquad (7.56)$$

for the critical temperature T_c, and to

$$j_\mathrm{c} = \lambda k_0 T_0^2 / Fd\{\phi - (e^3 F/\epsilon)^{\frac{1}{2}}\} \qquad (7.57)$$

for the current density at critical conditions. For thermal breakdown to occur, the current density (7.57) must be attained, but if the degree of field distortion required to produce avalanche breakdown occurs for smaller current densities, then thermal breakdown cannot occur. Examination of Fig. 7.10 shows that substantial distortion of the field occurs only if

$$JX^2 \geqslant 1. \tag{7.58}$$

Using eqns (5.36), (5.44), and (7.57) in (7.58), we get for the thickness marking the boundary between thermal and space-charge-enhanced critical conditions

$$d = \frac{F\mu_p\mu_n\epsilon H^3}{4\pi\lambda k_0 T_0^2 \alpha_n} \{\phi - (e^3 F/\epsilon)^{\frac{1}{2}}\}. \tag{7.59}$$

For thicknesses smaller than (7.59) breakdown should be thermal, and for larger thicknesses significant space charge may develop. To estimate a typical order of magnitude we use the dielectric parameters quoted in (5.48) together with

$$\left.\begin{array}{l} \lambda = 0\cdot1 \text{ J cm}^{-2} \text{ s}^{-1} \text{ K}^{-1} \\ \phi = 1\cdot0 \text{ eV} \\ T = 300 \text{ K} \\ F = 5 \text{ MV cm}^{-1} \end{array}\right\}. \tag{7.60}$$

(The value of λ is taken from Klein and Gaffni (1966).) These figures give $d \simeq 1000$ Å as the dividing line between the two types of breakdown. The equation (7.59) for the critical thickness is directly proportional to the product $\mu_n\mu_p$; unless this product is sufficiently small space charge cannot build up enough to cause substantial field distortion for any reasonable dielectric thickness.

To find an approximate formula for the critical field strength we must adopt some criterion for critical distortion of the field. Quite arbitrarily, it is assumed that

$$\mathscr{F}_{\text{cath}} > \mathscr{F} + 0\cdot2 \tag{7.61}$$

will correspond to critical conditions. It is possible to imagine many different types of conditions for critical field distortion, but since the objective is an approximate formula for the critical field strength, any convenient condition will suffice if it leads to results in agreement with the computations. Using (7.61) and the curves of Fig. 7.10, we find that critical conditions are well represented by

$$\mathscr{F}_c(J_cX^2)^{\frac{1}{3}} = 0\cdot5. \tag{7.62}$$

Using eqns (5.35), (5.36), and (5.44) to convert from dimensionless to actual quantities, and using eqns (7.55) and (7.61) in eqn (7.62), we obtain

$$(\bar{F}_c + 0\cdot2H)^{\frac{1}{2}} = \left(\frac{\epsilon}{e^3}\right)^{\frac{1}{2}}\left\{\phi - k_0 T \ln\left(\frac{32\pi\bar{F}_c^3 j_0\alpha_n d^2}{\mu_n\mu_p\epsilon H^6}\right)\right\}. \tag{7.63}$$

This is not an explicit result for the mean critical field strength, but it is in a form very suitable for successive approximations. From the data quoted in (5.48) and (7.60) together with a value of j_0 obtained from (3.52) and (3.53), it is found that eqn (7.63) reproduces the computed results of Fig. 7.9 to within a few per cent in the range 10^{-3} to 10^{-5} cm, and yields a result 25 per cent low for a thickness of 10^{-2} cm. The reason for the failure is a breakdown of the assumption that the current is entirely electronic; in fact, $x = 10^{-2}$ cm corresponds to $X = 100$ with the data of (5.48), and for this value of dimensionless thickness there is no region in which the current is predominantly electronic.

Various experimental investigations have supported the idea that breakdown is intimately tied in with the occurrence of a space charge, even though they do not at this time verify the details of the critical-field theory outlined above.

The distortion of fields in alkali halides was investigated by von Hippel, Gross, Jelatis, and Geller (1953) and Geller (1956) for KBr and by Williams (1964) for NaCl. These workers all used additively coloured crystals in which a positive space charge was created adjacent to the cathode by ionized F

centres which resulted from current flow consequent on photo-excitation. The field in this space-charge region was much higher than the mean field, and Williams found that for NaCl with a silver electrode noisy cathode-emission current set in when the field in the space-charge region reached a value of the order of the breakdown field strength. On the basis of this observation, he suggested that a contact phenomenon rather than an electron avalanche may be dominant in breakdown; the critical field discussed above actually envisages a contact phenomenon, which is caused by an electron avalanche, as being dominant.

Cooper and Elliott (1966) photographed the pre-breakdown light emission from KBr. They found that no light could be detected earlier than about 20 ns prior to collapse of voltage across the specimen, and thereafter a broad band of light emission composed of several bright cores propagated through the sample from the cathode to the anode. This evidence is considered to be unfavourable to the forty-generations theory and to the avalanche theory of Forlani and Minnaja (1964), since those theories both envisage the heaviest collision ionization as occurring in front of the anode, and the light emission would be expected to begin in that region. On the contrary, Cooper and Elliott (1966) believe that conditions first become unstable in the vicinity of the cathode, and that processes which enhance the field adjacent to the cathode must in effect produce a large increase in the current emitted from the cathode. Their work is thus a strong corroboration of the essential correctness of the approach to the critical field considered in this section.

A more detailed investigation of pre-breakdown light emission was done by Paracchini (1971), who measured the emission spectrum of NaI, KI, and RbI under the influence of a varying electric field. The alkali iodides were selected since their intrinsic recombination is radiative at liquid-nitrogen temperature, while for other alkali halides electrons and holes recombine through phonon processes down to considerably lower temperatures. Paracchini did in fact succeed in identifying the electrically stimulated emission as being due to the recombination of free electrons with V_K centres (these are essentially highly immobile holes in the alkali-halide valence bands). In another investigation using alternating voltage, Paracchini and Schianchi (1970) considered the role of injected carriers in the observed emission; since no electro-luminescence was observed with thin mylar films between the electrode and the alkali iodide, it was concluded that carrier injection is necessary for the occurrence of the emission. There are then two possibilities for the origin of the electrons and holes involved in the recombination process: double injection, i.e. electrons from the cathode and holes from the anode with consequent recombination, or single injection of electrons from the cathode with the holes generated by collision ionization. Paracchini and Schianchi (1970) reject the first hypothesis and accept the second one, since the extremely low mobility of the holes would prevent them from crossing to the cathode before being trapped. Similar conclusions were reached by Unger and Teegarden (1967); in no case however was the electro-luminescence observed with d.c. fields. This does not mean that d.c. fields cannot cause electro-luminescence (which would be in conflict with the findings of Cooper and Elliott (1966)), since the fields used in these experiments were an order of magnitude less than breakdown field strengths.

(b) Space charged produced by ionic migration

The suggestion that breakdown might be influenced by ionic space charges was made by Alger and von Hippel (1949) as a possible explanation of the sharp decrease in breakdown strength which they found for KBr above 200 °C (cf. Fig. 7.3); however, no convincing reasons appeared why the observed results could not be treated as impulse thermal breakdown. Cooper and Pulfrey (1971) also investigated KBr at 20 °C and −50 °C, and found that ionic migration through the crystal does not play a significant role in the breakdown process at these temperatures.

However, Watson and Heyes (1970) have interpreted breakdown results on NaCl at 300 °C and 350 °C as caused in part by ionic migration. Although 1:8000 μs impulses were used, the analysis for short times to breakdown can most simply be carried out by approximating the applied voltage as a square impulse. If positive ions are transported to the cathode at a rate that exceeds their discharge, then a positive space charge will form in front of the cathode, enhancing the field in this region. If a total space charge q is distributed uniformly over a distance x in front of the cathode, Poisson's equation gives

$$F_{\text{cath}} - F = 4\pi q / \epsilon_{\text{s}}, \tag{7.64}$$

where F is the field strength in the neutral body of the dielectric. The mean field strength will be given by

$$\langle F \rangle = F + (F_{\text{cath}} - F) x / 2d, \tag{7.65}$$

and combination of eqns (7.64) and (7.65) gives

$$F_{\text{cath}} = \langle F \rangle + \frac{4\pi q}{\epsilon_{\text{S}}} \left(1 - \frac{x}{2d} \right). \tag{7.66}$$

The rate of accumulation of charge at the cathode is assumed to be

$$dq/dt = \eta \sigma F, \tag{7.67}$$

where η is the fraction of charge transported that is not discharged at the electrode. If breakdown occurs within a short time of application of the impulse, the total space charge is small and F does not differ greatly from $\langle F \rangle$; under these circumstances a first-order solution to eqn (7.67) is

$$q = n\sigma \langle F \rangle t. \tag{7.68}$$

Substitution of eqn (7.68) in eqn (7.66) yields

$$F_{\text{cath}} = \langle F \rangle \left\{ 1 + \frac{4\pi \eta \sigma t}{\epsilon_{\text{S}}} \left(1 - \frac{x}{2d} \right) \right\}. \tag{7.69}$$

If we now assume that breakdown occurs when F_{cath} exceeds some assigned critical field strength F_{c}, the relation between the mean critical field $\langle F \rangle_{\text{c}}$ and the time is given by

$$\frac{F_{\text{c}}}{\langle F \rangle_{\text{c}}} = \left\{ 1 + \frac{4\pi \eta \sigma t}{\epsilon_{\text{S}}} \left(1 - \frac{x}{2d} \right) \right\}. \tag{7.70}$$

Watson and Heyes (1970) have compared their experimental results with this equation, as shown in Fig. 7.11. The assigned critical field F_c can be found from the zero time intercept, and the values are 2·5 MV cm⁻¹ and 2·7 MV cm⁻¹ at 300 °C and

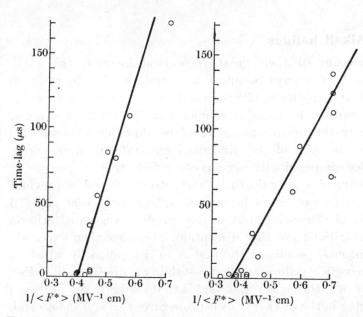

FIG. 7.11. Relationship between time-lags and the reciprocal of the mean field for square voltage impulses on NaCl at 300 °C and 350 °C (after Watson & Heyes 1970).

350 °C respectively. These fields are larger by a factor of two than commonly accepted values of the intrinsic critical field (cf. Fig. 7.3). If the ionic space-charge hypothesis is correct, they may correspond to critical fields for electron emission from the cathode. The values of η and $x/2d$ are not separately obtainable from this simple analysis; Watson and Heyes (1970) give a more elaborate analysis of breakdown occurring over longer time intervals, but find that in this region the evidence does not sufficiently distinguish between a space-charge and an impulse thermal mechanism.

BREAKDOWN IN SPECIAL SUBSTANCES

8.1. Alkali halides

ON account of their great theoretical interest, the alkali
halides have always occupied a central position in work on
dielectric breakdown, although it is not clear that the breakdown process in them is simpler than in other substances.
However they continue to be of considerable experimental
interest in spite of the difficulties encountered in preparing
samples compared with various other dielectrics.

The theories of critical field strength developed in previous
chapters do not go so far as to explain the actual physical
process of disruption, but merely produce suggested criteria
for the initiation of that disruption. There are then two main
experimental problems; the first is to determine by which of
the theoretical criteria the first instability is produced (if indeed
by any of them), and the second is to investigate the processes
occurring in the discharge as it propagates through the crystal.
While there is no clear-cut line marking off types of experiment
that determine the nature of the instability from those that
indicate the mechanism of the discharge, it remains broadly
true that investigations of the dependence of breakdown
strength on various parameters such as temperature, thickness,
or time of application of the applied voltage give the most
straightforward indications of the nature of the instability.

The most basic theoretical distinction of breakdown is that
between thermal and purely electrical. As already indicated,
the ratio of the electrical to the thermal conductivities is of
prime importance in determining the type of breakdown to be
expected. For the alkali halides in general the electrical conductivity follows a law of the form (1.2), while for the thermal
conductivity

$$\kappa = \kappa_0/T. \tag{8.1}$$

This means that purely electrical breakdown should be favoured for sufficiently low temperatures, and thermal breakdown for sufficiently high temperatures. In the transition from one type of mechanism to another there may well be a region of temperature in which experimental results depend in a very complex way upon the circumstances.

In the interest of clarity it seems reasonable to focus attention in the first place simply on the extent to which theory and experiment agree, only allowing the most rudimentary discussion of the justification of the theoretical assumptions. We shall follow this line in the present chapter; the whole situation will be reviewed in the following chapter with a view to examining in more detail the basis of the various theories.

In order to examine the experimental work in the light of these considerations, the dependence of breakdown strength on time of voltage application will be shown over a temperature range from liquid-air temperature up to some hundreds of degrees Celsius. The various regions of breakdown then separate out, and their assignment shows a very marked dependence on one further factor, the impurity concentration. The influence of other physical conditions such as the dielectric thickness, the cathode material, the crystal orientation, and heat treatment can then be discussed in the light of the fundamental mechanism operating. Changes in these factors would be expected to produce different results in regions of temperature in which different mechanisms of breakdown occur. Finally, it is of interest to study the variation of dielectric strength over the range of alkali halides, other factors remaining constant.

We commence discussion of results by noting a tacit assumption that the breakdown characteristics of the different alkali halides will be similar. This is a reasonable proposition in view of the similar physical properties of the alkali halides—notably that the form of temperature dependence of the electrical and thermal conductivities is the same for all of them. Figures 8.1, 8.2, and 8.3 show results of various workers for the breakdown strength of certain alkali halides as a function of temperature for various values of the rise time of the applied field.

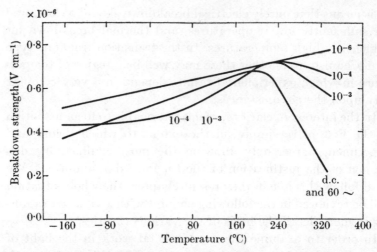

FIG. 8.1. The breakdown strength of KBr as a function of temperature for various rise times of the applied field (Alger and von Hippel 1949). No experimental points are given.

Comparison of the results of these diagrams shows the following main features.

(1) For sufficiently low temperatures the breakdown strength rises slowly with increasing temperature, and shows at most a slight dependence on the rise time of the applied field.

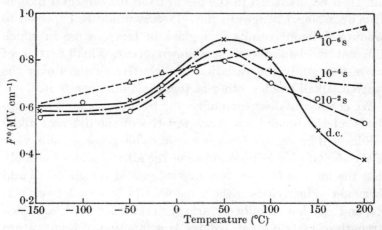

FIG. 8.2. The breakdown strength of KBr as a function of temperature for various rise times of the applied field (Konorova and Sorokina 1957).

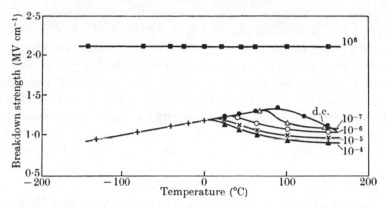

FIG. 8.3. The breakdown strength of NaCl as a function of temperature for various rise times of the applied field + All curves, ● d.c., ○ 100 μs impulses, × 10 μs impulses, ▲ 1 μs impulses, △ 0·1 μs impulses, ■ 0·01 μs impulses (Kuchin 1957.)

(2) At some temperature which is not sharply defined experimentally (but which is of the order of room temperature) the temperature dependence of the breakdown strength begins to depend on the rise time of the applied field. In addition, the situation in this region becomes confused because of the relatively greater disagreements of various workers.

We shall therefore divide discussion of possible breakdown mechanisms into three temperature regions. The low-temperature region extends up to the transition point noted above; the high-temperature region will be designated as that in which breakdown is thermal in nature, and the intermediate temperature region between these two. It should be mentioned that this division does not at any stage imply that definite assignable transition temperatures can be given for the different alkali halides; the boundaries of the regions may vary with many factors, but principally the degree of purity.

(a) *The temperature dependence of the breakdown strength at low temperatures*

The low-temperature breakdown of the alkali halides is believed to be electronic in nature; the main experimental support for this belief is the fact that breakdown strength is almost independent of the rise time of the field (cf. Figs. 8.1, 8.2, 8.3) over times from microseconds to minutes. In support of this contention, it seems hard to see how the power-dissipation process associated with any mechanism of current conduction (for a fixed number of carriers) can suddenly fail at some critical field under conditions only infinitesimally different from those associated with steady conditions. It is much more likely that failure of a steady state is associated with sudden multiplication of the number of current carriers, which thereby overcomes the power-dissipation process.

The intrinsic critical field strengths calculated in § 7.1 are all of the correct order of magnitude and all have a temperature dependence agreeing with that observed experimentally as shown in Fig. 7.3 for d.c. results. Although the understanding of breakdown has progressed a little further since the intrinsic critical-field theories were first elaborated, and no experimental worker would be content to label his observations of breakdown as intrinsic in the sense that the theories explained his results, nevertheless the theories probably have considerable significance as indicating a critical field strength at which collision ionization becomes large. In line with the theory and experimental work presented in § 7.3(a), breakdown in the alkali halides in the low-temperature region is probably due to electron emission from the cathode, enhanced by a large concentration of adjacent positive holes which are themselves the product of collision ionization. If this explanation is correct, the discharge current comes from the cathode, but the space charge that enables it to flow is caused by collision ionization, and hence massive collision ionization is the indirect rather than the direct cause of breakdown, and the temperature dependence of its onset is also the temperature dependence of the breakdown strength. The work of Cooper and Elliott (1966) and Paracchini (1971) discussed above lends particularly strong support to this hypothesis. Since Cooper and Elliott observed pre-breakdown light emission emanating from one or at most a few cores, one can only conclude that the space-charge distortion of the field occurs in a few channels between the electrodes rather than over the whole inter-electrode space.

(b) The breakdown strength at high temperatures

For sufficiently high temperatures the breakdown of alkali halides is undoubtedly thermal, and the experimental results have been compared with theory in §§ 6.2 and 6.3.

The lower temperature limit of the thermal region is difficult to determine since it depends on various experimental factors, but principally on the degree of specimen purity and on the manner of voltage application. The work of Hanscomb (1962) which was discussed in § 6.3 can be used to estimate the boundary of thermal breakdown. To do this we make the following assumptions.

(1) That the main temperature dependence of all thermal breakdown in substances obeying an electrical conductivity law (1.2) is given by

$$F^*_{\text{thermal}} \propto \exp(\phi/2k_0 T). \tag{8.2}$$

This is true for the special cases of steady-state and impulse thermal breakdown, and should hold for the intermediate case.

(2) That thermal breakdown field strengths cannot be expected to exceed about 1 MV cm^{-1} in the alkali halides. We assume in fact that some other mechanism would cause breakdown at about this field strength if thermal conditions were not yet critical.

It follows at once that the lower boundary of the thermal breakdown region depends on the rise time of the applied voltage and on the degree of purity of the sample. Taking Hanscomb's results and putting $\phi \sim 1$ eV, we have with the use of (8.1) estimated values for the lower boundary of the high-temperature region of NaCl as given in Table 8.1. A very high impurity concentration is required to shift the thermal boundary lower than about 200 °C, and since the other alkali halides have electrical and thermal conductivities of the same order of magnitude as those for NaCl similar boundary temperatures should apply for them.

TABLE 8.1

Estimated lower boundary temperature (°C) *for thermal breakdown of* NaCl.

Voltage rise time	Nominally pure Source 1	Source 2	Mole fraction $1\cdot25 \times 10^{-5}$	Mn^{++} impurity 8×10^{-4}
10 s	207	200	192	157
12 ms	283			
3·5 ms		300	283	222

It should be noted that drastic change in experimental circumstances could conceivably produce thermal breakdown at very low temperatures. Thin films set up in such a way as to have poor heat-dissipation properties could be expected to break down thermally as discussed in 7.3(*a*).

(c) *Breakdown at intermediate temperatures*

The intermediate temperature range extends roughly from room temperature up to 200 °C, and experimental work and its interpretation are beset with many difficulties in this area. The facts are reviewed by Hanscomb (1970) with particular reference to the effect of impurities and time of voltage application on the breakdown of KCl, for which the results are shown in Fig. 8.4. Basically his finding was that for short pulses (of order of tens of microseconds duration) the breakdown of KCl and KCl doped with $SrCl_2$ up to 300 °C appeared to be a continuation of the low-temperature mechanism. This conclusion is essentially in agreement with the result of Watson and Heyes (1970) on NaCl at 300 °C and 350 °C, but it does not entirely agree with results given by Cooper, Higgin, and Smith (1960) for KCl and KCl doped with $PbCl_2$.

Hanscomb rejects the ionic space-charge theory for the conditions of his experimental work, and seeks to explain the positive temperature characteristic shown in Fig. 8.4 by considering the reduction of electron mobility with increase of temperature. However, if the possibility of breakdown due to

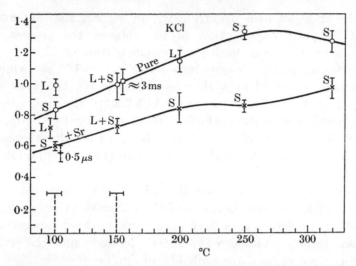

FIG. 8.4. Impulse breakdown strength of KCl as a function of temperature.
○ Nominally pure material; × 10^{-3} mole fraction $SrCl_2$ added. The notation
'L' refers to long-voltage impulses of 8 ms duration and S to short impulses
of 70 μs duration. (After Hanscomb 1970.)

space charge caused by positive holes is considered, then the
temperature dependence of hole mobility also plays a role in
determining the temperature dependence of the breakdown
strength. On physical grounds it is clear that increased hole
mobility causes increase in breakdown strength, since the
applied field is then more effective in preventing the accumulation of space charge. The approximate result (7.63) shows
the critical field to be weakly dependent on the hole mobility.
The decrease in breakdown strength that accompanies the
addition of impurity would also be explained on this hypothesis
if the impurity reduced the effective hole mobility; this would
occur, for example, if the divalent impurity—positive ion
vacancy complexes acted as traps for V_K centres.

The results of the various experimental workers may not be
in conflict to the extent that appears at first sight, when
account is taken of the different experimental details of their
work. The following hypotheses appear to be consistent with
the bulk of the experimental observations:

(1) For breakdown occurring within a short time interval ($\leqslant 10$ μs) after application of the voltage, the process is essentially continuous with the low-temperature one.

(2) For longer time intervals to breakdown (~ 100 μs), which have been observed at higher temperatures (cf. the work of Watson and Heyes (1970) quoted in § 7.3(b)), the ionic space charge adjacent to the cathode determines critical conditions.

(3) For still longer time intervals ($\geqslant 1$ ms), the thermal theory applies (cf. the work of Hanscomb (1962, 1969) quoted in § 6.3).

It should be borne in mind that these hypotheses are still oversimplified, chiefly because of the difficulty of accounting for all of the details that may play a role in this temperature region. Impulse-voltage wave form, electrode material, and specimen preparation are probably all significant.

(d) The influence of annealing specimens

The discussion of the preceding sections shows clearly that unrecognized factors have operated to prevent reproducibility in experimental results on the breakdown of alkali halides. Many attempts have been made to isolate these unknown factors, and, although a completely satisfactory situation has not been attained, nevertheless some significant points have been made.

The influence of annealing specimens prior to test was investigated by Calderwood, Cooper, and Wallace (1953), who presented a series of results on three groups of KCl crystals whose breakdown strength was measured at 20 °C. The first group underwent no heat treatment or mechanical stress other than that necessarily involved in their preparation, and the spread in breakdown strength extended from 0·54 to 1·54 MV cm^{-1}. The second group was annealed by being held near the melting point for 6 hours, and then allowed to cool gradually over a period of 12 hours; the spread for this group was reduced from 0·54 to 1·20 MV cm^{-1}. The third group was annealed and subsequently subjected to mechanical stress sufficient to cause some plastic deformation; a large spread was restored, from

TABLE 8.2

Effect of annealing and mechanical deformation on the breakdown strength of KCl *at* 20 °C (*After Calderwood, Cooper, and Wallace 1953*).

Specimen treatment	No. of specimens	Breakdown strength (MV cm⁻¹)	
		Mean	Variance
None	55	1·19	0·073
Annealing	43	1·01	0·051
Annealing and mechanical deformation	25	1·24	0·112

0·76 to 1·65 MV cm⁻¹. A statistical analysis of the results yielded the figures given in Table 8.2. Application of Student's *t* test shows that the mean of the annealed group is different from the mean of the whole population with better than 99 per cent confidence; furthermore, Fisher's *F* test gives the variance of the annealed and deformed group as different from the untreated group with better than 90 per cent confidence. The investigation leaves no doubt that heat treatment and mechanical deformation markedly influence the breakdown characteristics of KCl at 20 °C. Cooper (1962) also quotes similar results for KBr at 20 °C.

Hanscomb (1970) found that there was no appreciable influence of annealing on breakdown strength for KCl at 150 °C, and the existence of the effect discovered by Calderwood *et al.* therefore depends on the temperature. It is also conceivable that it may depend on the degree of purity of the alkali halide.

Cooper and Wallace (1953, 1956) examined the question as to whether it is possible, in principle, to obtain values of breakdown strength for undeformed alkali halides, since electrostatic forces will always stress the crystal. Using the existence of optical birefringence as a criterion for the existence of plastic deformation, they applied electric impulses of successively increasing amplitude to previously annealed crystals. It was found that birefringence occurred prior to breakdown in about half of the samples of KCl and NaCl tested, but that no plastic deformation was observed in annealed specimens of KBr.

(e) *The influence of impurities*

Substitutional divalent impurities dominate the situation in the high-temperature region, since they determine the magnitude and the temperature dependence of the electrical conductivity, which in turn determines the thermal breakdown strength. We shall be concerned in this section with the possibility of smaller effects caused by substitutional impurities in the low- or intermediate-temperature regions.

It seems logical to check first on the effect of substitutional impurities of other alkalis or halides, which should presumably be very common. The breakdown strengths of a series of mixed KCl–RbCl crystals were measured by von Hippel (1934), whose results are shown in Fig. 8.5. Measurements on mixed KCl–KBr yielded similar results; if this similarity holds in other cases, then small concentrations of such impurity will increase the breakdown strength.

The effect of metallic impurity ions that are chemically less similar than another alkali has been investigated by von Hippel and Lee (1941), who added AgCl to NaCl, Cooper *et al.* (1960) who added $PbCl_2$ to KCl and $CdCl_2$ to NaCl, and also by Hanscomb (1970) who added $SrCl_2$ to KCl and $MnCl_2$ to NaCl.

The results of von Hippel and Lee (1941) are shown in Fig. 8.6, from which it is apparent that the admixture of silver very greatly raises the breakdown strength in the low-temperature region; at intermediate temperatures the effect is reversed after a complicated cross-over. The impurity in these experiments has been added in very large amounts, but even so there remains room for suspicion that, in d.c. tests of long duration, the migration of metal ions into NaCl from a silver anode may substantially alter the breakdown strength (cf. Harris 1968). In this connection it should be noted that Cooper *et al.* (1960) added smaller quantities (0·02 mole per cent) of AgCl to KCl

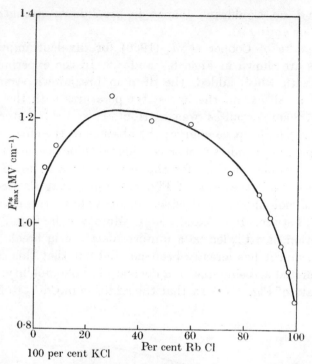

FIG. 8.5. The breakdown strength of mixed crystals of KCl–RbCl at 20 °C
(after von Hippel 1934).

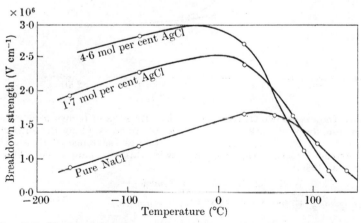

FIG. 8.6. The breakdown strength of NaCl as a function of temperature with
various admixtures of AgCl (after von Hippel and Lee 1941).

and found no significant change in breakdown strength at
−195 °C or at 100 °C.

The results of Cooper *et al.* (1960) for divalent impurity
additions are shown in Figs. 8.7 and 8.8. In the experiments
on KCl with $PbCl_2$ added, the drop in breakdown strength
observed at all except the lowest temperatures with the flat-
topped 1:8000 μs pulses was attributed to field distortion
arising from ionic space charge; the absence of the effect with
the sharp 0·5:5 μs pulses is then considered to be because there
has been insufficient time for the space charge to build up.
Comparison with the data of Fig. 8.4 shows that Hanscomb
(1970) did not observe a decrease in breakdown strength for
pure KCl between 100 °C and 200 °C. Moreover, he found that
the addition of $SrCl_2$ led to a uniform decrease in breakdown
strength, and it has already been pointed out that this could
be interpreted as being due to a decrease in hole mobility.

The data of Fig. 8.8 show that the addition of $CdCl_2$ to NaCl

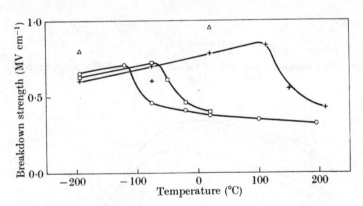

Fig. 8.7. The breakdown strength of KCl as a function of temperature with
various admixtures of $PbCl_2$. The curves shown were drawn by the authors
through the minimum measurements. In all cases the range of measured
values extended to somewhat less than twice this minimum value. (After
Cooper *et al.* 1960.)

+ Nominally pure   ⎫ 1:8000 μs
☐ 0·009 mole % $PbCl_2$ ⎬   impulses.
○ 0·022 mole % $PbCl_2$ ⎭
△ 0·07 mole % $PbCl_2$ 0·5:5 μs impulses.

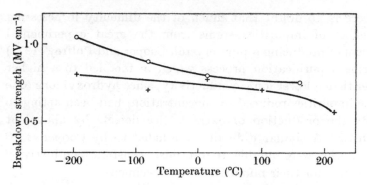

FIG. 8.8. The breakdown strength of NaCl as a function of temperature with various admixtures of $CdCl_2$. The curves are drawn as in Fig. 8.7.

+ Nominally pure⎫ 1:8000 µs

◯ 0·096 mole % ⎬ impulses.

(After Cooper *et al.* 1960.)

causes an increase in breakdown strength at all temperatures measured. This conclusion is supported qualitatively by Hanscomb (1970), whose data for NaCl with $MnCl_2$ added are shown in Table 8.3. No convincing explanation has been proposed for this impurity-induced rise in the breakdown strength.

TABLE 8.3

Breakdown strength of NaCl *both nominally pure and with* 10^{-3} *mole fraction* $MnCl_2$; *the notation* L *refers to long voltage impulses of* 8 ms *duration and* S *to short impulses of* 70 µs *duration* (*After Hanscomb 1970*).

Temperature (°C)	Crystal	Voltage pulse type	Mean (MV cm^{-1})	5% confidence interval (MV cm^{-1})
100	Pure	L & S	1·16	1·10–1·22
	Doped	L & S	1·37	1·28–1·49
200	Pure	S	1·30	1·23–1·36
	Doped	S	1·51	1·42–1·61

There is no doubt that much of the difficulty in assessing the effect of impurities stems from the great experimental problem of producing a pure crystal. Cooper and Pulfrey (1971) describe a purification process which in fact led to a higher rather than a lower ionic conductivity, since hydroxyl ions (one of the impurities reduced in concentration) had been acting to nullify the production of extra lattice defects by aliovalent impurities. A similar difficulty is alluded to by Cooper *et al.* (1960), who were unable to establish constant conductivity parameters for their nominally pure specimens.

(f) Directional effects in breakdown

There are two principal directional effects possible in alkali halides, the discussions of which greatly overlap. In the first place, the magnitude of the breakdown strength may depend on the crystallographic direction in which the field is applied, and secondly there may be preferred directions for the formation of the discharge paths.

The second of these effects was observed by Inge and Walther (1930), von Hippel (1931), and Lass (1931), using an inhomogeneous applied field at room temperature. More comprehensive studies, which yield the preferred direction of discharge path as a function of temperature, were performed by Davisson (1946, 1959), Caspari (1955), and Cooper *et al.* (1957). The results of this work are shown in Tables 8.4 and 8.5 for NaCl and KCl respectively. There have been two explanations of these observations, the one based on a further series of experiments and the other theoretical.

Cooper *et al.* (1957) argue that the alkali halides must be anisotropic with respect to breakdown strength, if the discharge tracks originating from a point electrode producing a spherically symmetric field system follow definite crystallographic directions. In order to test this hypothesis, they measured the breakdown strengths of NaCl and KCl in a uniform field for three principal crystallographic directions, and over a wide range of temperatures. Their results are shown in Figs. 8.9 and 8.10. These results can be used to predict preferred directions

TABLE 8.4

Preferred crystallographic directions of breakdown paths in NaCl. In some cases different directions are preferred; subscripts indicate the relative preferences where these are measured.

Temperature	Observed by			Predicted by, Cooper, Grossart, Wallace	Calculated by Callen, Offenbacher
	Davisson	Caspari	Cooper, Grossart, Wallace		
−200 °C	(100) Random	(111) (110)	$(100)_{10}$ $(110)_{1}$	(100)	(100) Random
−74 °C	—	110	$(110)_{4}$ $(100)_{1}$	(100) (110)	(100) Random
20 °C	(110) (111)	(110)	$(110)_{4}$ $(100)_{1}$	Random	100
115 °C	(110)	(110)	$(110)_{15}$ $(100)_{5}$ $(111)_{1}$	(110) (100)	(110)
200 °C	(100)	(110)	$(110)_{4}$ $(100)_{1}$	Random	(100)
400 °C	(110)				(100)

TABLE 8.5

Preferred crystallographic directions of breakdown paths in KCl.

Temperature	Observed by Davisson	Observed by Cooper, Grossart, Wallace	Predicted by Cooper, Grossart, Wallace	Calculated by Callen, Offenbacher
−200 °C	(100)	(100) (110)	(100)	(100) Random
−74 °C		(100)	(100)	(111)
20 °C	(100)	(100)	(100)	(100) (110)
200 °C	(100)	(100)	Random	(100)
400 °C	(100) (110)	—	—	(100)

of the discharge path as stated, and these directions are entered in Tables 8.4 and 8.5.

Cooper and Fernadez (1958) have found that the direction of an applied uniform field and the direction of the discharge path are sometimes different. In a large number of measurements on KCl at room temperature with the field applied in the 110 crystallographic direction, they observed discharge tracks in the 100 direction for those specimens that had breakdown strengths among the lower values, and 110 discharge tracks for the specimens of higher breakdown strength. This is further convincing evidence of the marked anisotropy of KCl at room temperature; in fact, it appears to be markedly anisotropic from the lowest temperatures up to 115 °C. On the other hand, NaCl is perhaps anisotropic at −195 °C and approximately isotropic at all higher temperatures.

Cooper and Elliott (1968) observed pre-breakdown light emission in KCl for fields applied in various crystallographic directions; they found that the discharge channel does follow the same path as that indicated by the bright core of light emission. This evidence for the existence of electronic instabilities in directions other than that of the applied field appears to rule out hot-electron theories of breakdown. It was

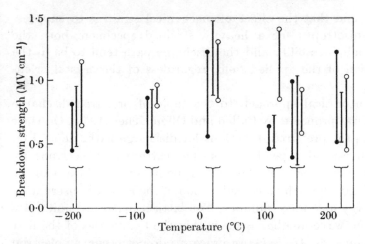

FIG. 8.9. The breakdown strength of KCl in a uniform field as a function of temperature and crystallographic direction with 1:50 μs impulses. The vertical lines indicate the spread of measured values; in all cases the number of specimens tested was of order of 20. ⋮ (100) specimens. I (110) specimens. ⅋ (111) specimens. (After Cooper *et al.* 1957.)

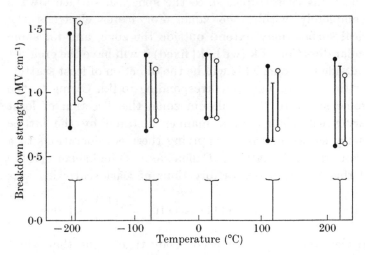

FIG. 8.10. The breakdown strength of NaCl in a uniform field as a function of temperature and crystallographic direction with 1:5000 μs impulses. The vertical lines indicate the spread of measured values; in all cases the number of specimens tested was of order of 20. ⋮ (100) specimens. I (110) specimens. ⅋ (111) specimens. (After Cooper *et al.* 1957.)

also found that the effect of mechanical strain is to suppress any anisotropy; for a heavily strained specimen, both the electronic instability and the discharge path tend to be in the direction of the applied field, regardless of the crystal orientation.

A theoretical approach to the topic of preferred discharge paths was proposed by Callen and Offenbacher (1953). On their theory, the preferred directions for discharge paths depend on the directional dependence of the scattering of electrons by lattice vibrations. The essential point of the idea can be explained simply by a consideration of the laws of conservation of energy and wave number. The lattice modes are uniformly dense in wave-number space within the boundaries of the first Brillouin zone. If *Umklapp* processes do not occur, an electron of wave vector $\mathbf{k}$ can interact only with those lattice modes that lie approximately on a spherical surface of radius $|\mathbf{k}|$, and whose centre is at the point $-\mathbf{k}$ relative to the centre of the Brillouin zone (cf. eqns (2.38) and (2.49)). The scattering of slow electrons is isotropic, since the spherical surface always lies completely within the zone. For faster electrons the spherical surface may extend outside the zone, and for some particular direction of $\mathbf{k}$ (with $|\mathbf{k}|$ fixed) it will have the greatest area outside the zone. This will be the direction of least scattering for the electron energy corresponding to $|\mathbf{k}|$. Owing to the polygonal shape of the Brillouin zone, the direction of least scattering will change in a manner dictated by the lattice symmetry as $|\mathbf{k}|$ increases. Applying these considerations to a face-centred cubic lattice, Callen and Offenbacher (1953) found that the sequence of directions of least scattering was

$$\text{Random} \rightarrow (100) \rightarrow (111) \rightarrow (110) \rightarrow \frac{(110)}{(100)} \rightarrow (100)$$

as the electron energy increased. They then argue that, since the thermal barrier to the electron motion moves to higher energies with increasing temperature, the sequence of directions of least scattering with increasing electron energy is the same as the sequence of directions of preferred discharge track with increasing temperature. The sequence is listed alongside the experimental results in Tables 8.4 and 8.5; the extent of agreement is only moderate. Considerations such as these retain their validity whether one is thinking of an intrinsic critical field or of a critical field explained in terms of electron space-charge-enhanced cathode emission. In the latter case, the immobile holes that constitute the space charge and cause the enhanced cathode emission lie along directions of easy collision ionization, and this is essentially what has been calculated.

(g) The effect of thickness on breakdown strength

A knowledge of the effect of thickness on the breakdown strength can be of considerable importance in determining the breakdown mechanism. Thus, for example, impulse thermal breakdown strength should be thickness-independent, while other forms of thermal breakdown depend on thickness either strongly or weakly, as indicated in Chapter 6. Intrinsic critical field strengths are thickness-independent, but if they are regarded as criteria determining the onset of substantial collision ionization rather than as criteria for breakdown itself, this will not be reflected in the thickness dependence of breakdown strengths.

The various avalanche and space-charge-enhanced emission theories all depend weakly on thickness, and therefore require comprehensive experimental data to check their validity. The most extensive results available are those for NaCl shown in Fig. 7.5 and representing a compilation of the results of different experimental groups. The breakdown data extend over almost three decades of specimen thickness and can be well fitted by the forty-generations avalanche theory or by the space-charge-enhanced cathode-emission theory; it has been pointed out that the latter is to be preferred since it does not lead to impossibly high collision-ionization coefficients.

Other thickness-dependent measurements are available, most of which cover a range of one decade or less. Schissler (1960) investigated breakdown in KBr with 1 μs impulses at −150 °C,

20 °C, and 100 °C over a range of thickness. His experimental points and the curves which he constructed are shown in Fig. 8.11. The temperature dependence of the thickness effect could presumably be accounted for by the space-charge-enhanced cathode-emission theory, which is represented approximately by eqn (7.63). However, in view of the uncertainty in the value of many quantities in this equation, it does not seem possible to determine the exact role of the temperature in the thickness effect.

Vorob'ev, Vorob'ev, and Murashko (1963) measured the thickness effect in NaCl for thin specimens (5–20 μm) at room temperature, and for different crystallographic directions. Their results are shown in Fig. 8.12. It is noteworthy that these authors claim a directional effect for the small thickness in question (cf. the results of Cooper et al. (1957) who found NaCl to be isotropic at room temperature, but who used thicker specimens).

Smith and Budenstein (1969) found an approximate fit to eqn (7.46) for NaF and LiF over a thickness range 10^{-5} to 10^{-4} cm. However, as pointed out above, a good deal more than one decade of thickness would be required to establish conclusively the inverse half-power thickness dependence. Other measurements of thickness dependence in thin films of alkali halides have been done by Plessner (1948) on NaF and KBr, and by Weaver and Macleod (1965) on NaCl and LiF.

All the evidence for alkali halides points unambiguously to an increase of breakdown strength with decreasing thickness. Such a dependence is predicted by most of the theories of avalanche critical fields and space-charge-enhanced-emission critical fields. None of these theories could be said to offer an adequate explanation of the facts, but the discussion in § 7.3(a) is probably the most soundly based.

(h) Electrode effects and time-lags

The possible effect on the breakdown strength of different electrode materials depends on the temperature region under consideration. In the high-temperature region, the thermal

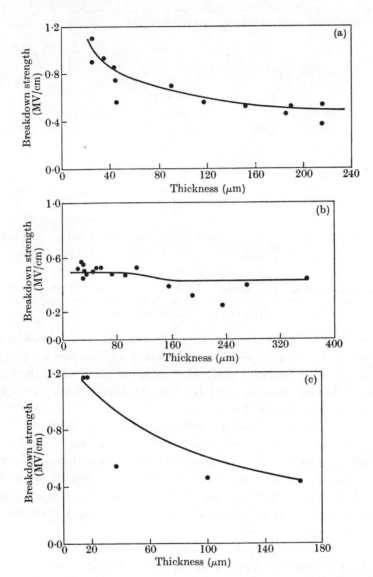

Fig. 8.11. The breakdown strength of KBr as a function of thickness for 1 μs impulses and various temperatures: (a) 20 °C, (b) −150 °C, (c) 100 °C. (After Schissler 1960.)

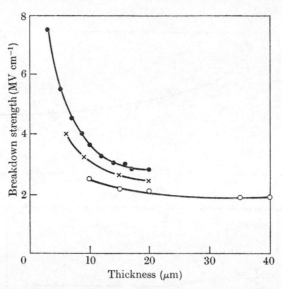

FIG. 8.12. The breakdown strength of NaCl as a function of thickness at room temperature for various crystallographic directions. ● (100) direction, × (111) direction, ○ (110) direction. (After Vorob'ev *et al.* 1963.)

properties of the electrode system influence the breakdown in a way that is, in principle, well-understood (cf. Chapter 6); there remains the question as to whether different electrode materials may influence breakdown in the low- or intermediate- temperature regions. Briefly, the experimental answer to this question is that different metallic electrode materials yield slightly different breakdown strengths, but that use of an electrolytic solution as cathode produces a markedly different result from that obtained with a metallic cathode. The question of time-lags in the breakdown process has been included in this section, since results on it have been found to be strongly correlated with electrode material.

Alger and von Hippel (1949) found that the material of the cathode markedly changed the room-temperature d.c. breakdown strength of KBr. The electrode materials used were ball-bearings, gold film, mercury, and KBr solution. For cases in which spurious discharges were eliminated, the average breakdown strength varied from 0·54 to 0·88 MV cm⁻¹; the test

TABLE 8.6

The effect of pre-treatment and electrode material on the d.c. breakdown strength of KBr at room temperature (von Hippel and Alger 1949).

Electrode system	Pre-treatment	Average breakdown strength	Remarks
Ball-bearings without guard	None	0·38	Low figure presumably due to discharges in the ambient
Ball-bearings with wax guard	None	0·88	The specimen had presumably been exposed to atmospheric moisture
Gold film	Heated to 375 °C in vacuum	0·54	
Mercury	Heated to 375 °C in vacuum	0·54	
KBr solution	None	0·84	
Gold film	Stored in moist atmosphere	0·79	

results are summarized in Table 8.6. However, the specimens used did not have uniform treatment otherwise; some were heated to 375 °C in vacuum before breakdown, while others were stored in moist conditions. Since low breakdown strengths were correlated with previous heating and high ones with moist conditions, the significance of the cathode material is somewhat uncertain. However, Alger and von Hippel (1949) do claim (without giving figures) that gold-cathode–liquid-anode breakdown results were lower than the values when the poles were reversed.

Cooper and Grossart (1953) measured the d.c. and impulse breakdown strength of KBr at room temperature using cathodes of graphite, gold, or KBr solution. There was no significant difference between the mean breakdown strengths of the various groups. Similar measurements were carried out by Calderwood *et al.* (1953) on KCl. Using d.c. at room temperature, they found no significant difference between mean breakdown strengths for graphite and silver electrodes. However, electrolytic cathodes gave a higher mean breakdown strength than graphite cathodes, as shown in the results of Table 8.7; the means are different with better than 98 per cent confidence.

TABLE 8.7

The effect of different electrodes on the breakdown strength of KCl *at* 20 °C (*Calderwood et al. 1953*).

Electrode	No. of specimens	Breakdown strength Mean (MV cm^{-1})	Standard deviation (MV cm^{-1})
Graphite anode-graphite cathode	13	0·93	0·213
Graphite anode-electrolytic cathode	10	1·17	0·249

In a very large series of experiments on the breakdown of annealed specimens of NaCl, Cooper and Smith (1961) produced the results shown in Table 8.8 for the effect of different surface treatments. Statistical analysis of the results shows that the mean breakdown strength with gold electrodes is significantly different from that with either sodium or silver, with 99·5 per cent confidence. Since the metallic electrodes were applied to polished surfaces and 4 per cent voltage increments were applied, the results for graphite electrodes under the same conditions should be used. Calculation shows that there is not a statistically significant difference between the results for graphite and for any metal electrode.

On account of the large number of specimens tested, the results of Cooper and his co-workers are highly significant. They show that gold electrodes lead to breakdown strengths of

TABLE 8.8

The influence of the cathode material on the breakdown strength of NaCl at 20 °C with 0·8:8000 μs voltage impulses (Cooper and Smith 1961).

Cathode material	Surface treatment	Percentage voltage increment	No. of specimens	Breakdown strength	
				Mean (MV cm^{-1})	Standard deviation (MV cm^{-1})
Graphite	Polished	4	37	1·06	0·137
Graphite	Not polished	4	73	1·01	0·135
Sodium	Polished	4	68	0·99	0·201
Gold	Polished	4	29	1·12	0·204
Silver	Polished	4	33	0·98	0·158
Graphite	Polished	1	31	0·98	0·099
Graphite	Not polished	12	32	0·98	0·103

the order of 10 per cent higher than those for other metals, and that electrolytic solutions as cathode give results of the order of 25 per cent higher. These facts find a qualitative explanation on the basis of the space-charge-enhanced cathode-emission theory, since the cathodes with the highest work functions lead to the highest breakdown strengths.

We turn now to the question of time-lags in the breakdown of alkali halides. This topic has been discussed in § 7.3(b) in connection with the work of Watson and Heyes (1970) on NaCl at 300 °C and 350 °C; those authors interpreted the time-lag between the application of the field and breakdown as the time required for an ionic space charge to form in front of the cathode. A completely different approach was taken by Cooper and Smith (1961), who investigated time-lags in the breakdown of NaCl at room temperature. They assumed that breakdown was caused by a critical avalanche of electrons, and the time-lag was therefore interpreted statistically.

The discussion of statistical time-lag is conveniently given in terms of the single-electron avalanche mechanism of § 7.2(b). (It is not necessarily implied that the validity of the statistical time-lag concept is tied to the forty-generations avalanche theory; it could apply equally well to a space-charge-enhanced emission theory, since the space charge is caused by an electron avalanche.) Let n_0 be the number of electrons in a critical avalanche, and let $\bar{n}$ be the average avalanche under the given conditions of specimen thickness and applied field. It is readily shown that the probability that an avalanche will be greater than the critical is $\exp(-n_0/\bar{n})$ (cf. Wijsman 1949). Hence, if there are ν_0 avalanches initiated per unit time, then the mean time elapsed before a critical avalanche is

$$\tau = \nu_0^{-1} \exp(n_0/\bar{n}), \qquad (8.3)$$

which is called the mean statistical time-lag. Recalling that $\bar{n} = \exp\{\alpha(F)d\}$, where d is the specimen thickness and $\alpha(F)$ the mean rate of ionization per unit length, we have

$$\tau = \nu_0^{-1} \exp[n_0/\exp\{\alpha(F)d\}], \qquad (8.4)$$

which shows a very strong dependence on specimen thickness and applied field.

Consider now a series of N_0 experiments in which a given field is applied to specimens of the same thickness. This will result in $N(t)$ cases in which breakdown does not take place before the time t. As the probability of the occurrence of a critical avalanche in an ensuing interval of time does not depend on the time t since application of the voltage, we have

$$N(t) = N_0 \exp(-t/\tau) \qquad (8.5)$$

for the distribution of observed time-lags.

Cooper and Smith (1961) fitted the results of a very large number of experiments on NaCl to eqn (8.6) to find the statistical time-lag. The experiments consisted in applying a series of $0{\cdot}8{:}8000$ μs impulses (the amplitude of each impulse exceeding the previous one by the same percentage) and measuring the time-lag from the peak of the wave until the collapse at breakdown. This time is the statistical time-lag and is identical with the time-lag that, at a much higher temperature, Watson and Heyes (1970) attribute to the formation of an ionic space charge. The formative time-lag (i.e. the time occupied in the actual voltage collapse) was estimated by Cooper and Smith to be less than 4×10^{-8} s. Using improved electronic circuitry, Bradwell and Pulfrey (1968) confirmed that this figure is conservative by measuring formative time-lags of 1 to 5×10^{-9} s in KBr. The effect of the cathode material on the mean statistical time-lag is given in Table 8.9 for groups of annealed specimens of NaCl of varilus thicknesses and with 4 per cent voltage increments between successive impulses. Referring to eqn (8.4), we see that the mean statistical time-lag should depend on the rate of initiation of avalanches (which is presumably dependent on the cathode material) and on the specimen thickness. Assuming that specimen thickness and electrode material were uncorrelated, we see from Table 8.9 that surface treatment and electrode material have a very marked influence on the value of the mean statistical time-lag.

TABLE 8.9

Mean statistical time-lag as a function of electrode material and surface treatment of NaCl *using* 4 *per cent voltage increments and for specimens* 0·01 cm *thick* (*Cooper and Smith 1961*).

Cathode	Surface treatment	No. of specimens	Mean statistical time-lag (μs)
Sodium	Polished	67	1·0
Graphite	Polished	100	3·2
Gold	Polished		3
Silver	Polished		3
Graphite	Unpolished	66	11

The dependence of the mean statistical time-lag on the applied field has been studied by Cooper and Smith (1961) by varying the percentage voltage increment between successive impulses applied to the specimen. Their results are shown in Table 8.10, which in a certain manner measures the effect of 'overvoltages'. It has been pointed out by Cooper (1962) that the use of this term is not without objection in the current context, since it is impossible to determine the degree of overvoltage without destructive breakdown; however, for practical purposes it can be taken that the average breakdown strength is to be found from results on long pulses. Using this notion of overvoltage, Kawamura, Ohkura, and Kikuchi (1954)

TABLE 8.10

Mean statistical time-lag as a function of percentage voltage increment for graphite electrodes applied to polished surfaces of NaCl *at room temperature* (*Cooper and Smith 1961*).

Percentage voltage increment	No. of specimens	Mean statistical time-lag (μs)
1	29	17
4	95	4·5
12		0·8 (estimated)

measured the mean statistical time-lag for KCl at room temperature; they give a result of 0·24 μs at 8 per cent overvoltage and <0·03 μs at 18 per cent overvoltage. Since Cooper and Smith worked with NaCl, and introduced overvoltages in a different way, the results are not strictly comparable; however, the measured time-lags are of the same order of magnitude under similar conditions.

(i) The alkali halides compared

Measurements on the d.c. breakdown strength of various alkali halides at room temperature were made by von Hippel (1935a) whose results are shown in Table 8.11. There have been several attempts to find an empirical equation which relates these measured breakdown strengths to some other crystal parameter. Franz (1956) gives for the d.c. breakdown strength at room temperature

$$F^* \simeq 26/a^3 \qquad (8.6)$$

where F^* is in MV cm^{-1} and a is the interionic distance in Å.

TABLE 8.11

Comparison of various field strengths in MV cm^{-1} for the alkali halides

Substance	Room-temperature d.c. breakdown strength (von Hippel 1935a)	Field F_0 of eqn (7.21)	CO$_2$ laser-induced breakdown strength (Yablonovitch 1971)
LiF	3·05	6·2	
NaF	2·40	4·4	
NaCl	1·50	1·61	1·95
NaBr	0·83	1·01	0·91
NaI	0·69	0·82	0·79
KF	1·80	3·8	2·40
KCl	1·00	1·26	1·39
KBr	0·69	0·83	0·94
KI	0·57	0·53	0·72
RbCl	0·83	1·08	0·93
RbBr	0·58	0·69	0·78
RbI	0·49	0·45	0·63

Vorob'ev (1956) gives
$$F^* \simeq 0{\cdot}6E_c - 3 \qquad (8.7)$$

where E_c is the cohesive energy in eV per ion pair.

Theoretically, the Fröhlich–Paranjape (1956) intrinsic critical field gives the simplest comparative results. Their critical field differs by a factor of order unity from the field strength F_0 of eqn (7.21); this latter field is in fact approximately 1·5 times the critical field at the Debye temperature. A list of values of F_0 for various alkali halides is given in Table 8.11; as pointed out in § 8.1(f), there are serious objections to the hot-electron intrinsic critical-field theory, but F_0 is a characteristic field strength, expressed in simple terms, which presumably bears some relation to the onset of collision ionization.

Yablonovitch (1971) has studied CO_2 laser-induced breakdown in various alkali halides, and his results are shown in Table 8.11. If it is assumed that energy loss by electrons to the lattice is governed by a Debye-type factor $(1+\omega^2\tau^2)^{\frac{1}{2}}$ where ω is the frequency of the applied field, it is reasonable to assume further that
$$F_{rms}^*(\omega) = (1+\omega^2\tau^2)^{\frac{1}{2}}F_{dc}^* \qquad (8.8)$$

for any breakdown process that depends on collision ionization. For the CO_2 laser, $\omega \sim 10^{14}$ s^{-1}; eqn (2.74) yields $\tau \sim 10^{-14}$ s for a typical alkali halide at room temperature. Since $\omega\tau \sim 1$, the CO_2 laser-induced breakdown should be of the order of 50 per cent greater than the d.c. breakdown strength. This is borne out by Yablonovitch's results.

The correct trend from one substance to another, and the comparison between magnitudes when taken along with the other evidence, leave little room for doubt that breakdown in the alkali halides is at least initiated by an electron avalanche, and that the mechanism tending to inhibit the formation of the avalanche is interaction with lattice vibrations.

8.2. Glass and quartz

Materials of technical importance were investigated by electrical engineers long before work had commenced on alkali halides. Most of the early work on glass was inconsistent and uncertain and will not be presented here (for a bibliography and review, see Vermeer (1959)). It is convenient to divide the discussion into high-field conduction, breakdown, and properties of SiO_2 films.

(a) High-field conduction

The intensive data of Vermeer (1956a) on high-field conduction in various glasses have been shown in Fig. 4.2, and it was noted that Vermeer fitted his results to the Mott and Gurney law (4.1) for high-field ionic mobility. Barton (1970) has investigated the similarity between the electronic conduction in glasses containing transition-metal ions and the ionic conduction in glasses containing alkali. His results are shown on a Schottky plot in Fig. 8.13 for two glasses, one of which conducts ionically and the other electronically. Since

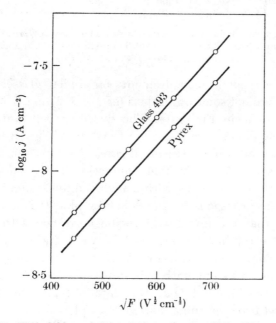

FIG. 8.13. High-field conduction in two glasses. Glass 493 contains 8 mole % of Fe_2O_3, the specimen thickness was 103 μm, and the temperature $-10\ °C$. Pyrex glass contains 3·5 mole % of Na_2O, the specimen thickness was 189 μm and the temperature 66 °C. (After Barton 1970.)

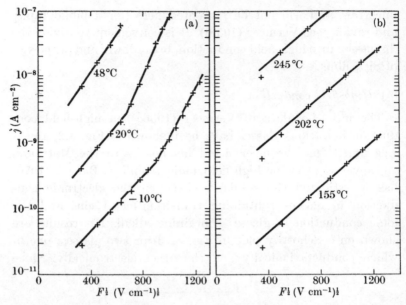

FIG. 8.14. High-field conduction in two glasses. The data of Vermeer (Fig. 4.2) replotted on a Schottky diagram: (a) Thuringian glass, (b) Phillips 18 glass.

Vermeer (1956a) made measurements on similar glasses, it is of interest to present some of his data (cf. Fig. 4.2) on a Schottky plot; this is done in Fig. 8.14. The dielectric constants and activation energies determined from the data of Figs. 8.13 and 8.14 are shown in Table 8.12; in all cases the data found refer to uncompensated Poole–Frenkel conduction (cf. § 4.3(b)).

Vermeer's results on glass with a high sodium content show two distinct linear regions on the Schottky plot. It is tempting to postulate that the lower-field region corresponds to field-enhanced ionic conduction (cf. § 4.1(b)) while the higher-field region corresponds to a Poole–Frenkel type effect. The data of Table 8.12 are consistent with this interpretation. However it should be borne in mind that the data obtained by fitting Poole–Frenkel type formulae to Schottky plots are not highly reliable; the activation energy is not accurately determined, and the meaning of the value for the dielectric constant is diminished by conceptual difficulties with the model.

TABLE 8.12

Parameters for Poole–Frenkel conduction (uncompensated) determined from the data of Figs. 8.13 and 8.14. The values of the dielectric constant differ from those given by Barton (1970) since he used the result appropriate to a compensated material.

Glass	Activation energy (eV)	Dielectric constant
Pyrex	1·03	4·1
Glass 493	0·62	6·1
Phillips 18	1·19	7·2
Thuringian (medium field)	0·66	12·2
Thuringian (high field)	0·71	3·8

(b) Breakdown

The temperature dependence of the breakdown strengths of soda-lime glass, silica glass (fused quartz), and crystalline quartz was measured by von Hippel and Maueer (1941) using d.c. voltages. For the glasses they found that the breakdown strength was nearly constant from liquid-air temperature up to room temperature, but that it fell very steeply thereafter. For the crystalline quartz the temperature characteristic rose throughout the range; the results for crystalline and fused quartz are shown in Fig. 8.15. The breakdown strength of crystalline quartz has the same positive temperature characteristic as the alkali halides, and the breakdown mechanism is probably similar. Quartz has several infrared frequencies, and optical experiments show two particularly strong reflections at about 10 and 20 μm wavelength (cf. Lecomte 1958).

Since the highest frequency will give an upper limit to the characteristic field strength F_0 of eqn (7.21), we take $\omega_t \simeq 2 \times 10^{14}$ s^{-1}, and using $\epsilon_s \simeq 4\cdot3$ and $\epsilon_m \simeq 2\cdot3$ we find that $F_0 \simeq 6\cdot7$ MV cm^{-1}. This is of a reasonable order of magnitude;

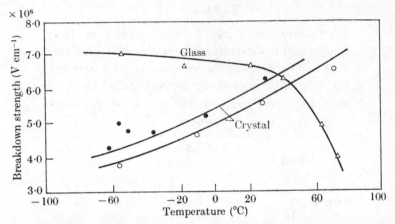

Fig. 8.15. The d.c. breakdown strengths of fused and crystalline quartz as a function of temperature (after von Hippel and Maurer 1941).

if the breakdown is due to space-charge-enhanced cathode emission, the exact value of breakdown strength will depend in a complex way on many factors not included in the characteristic field strength F_0. The strength of fused quartz, after remaining almost constant up to room temperature, then falls sharply; this was interpreted by Fröhlich (1947a) in terms of the intrinsic critical field strength of eqn (4.65). Tomura and Kikuchi (1952) measured the breakdown strength and conductivity of a soda-lime glass using flat-topped pulses of 30 and 110 μs duration over the temperature range 170 °C to 220 °C. They confirmed the form (4.65) for the breakdown strength, and also found that the high-field conductivity followed a law of the form (4.68), which follows from the same model. However, the critical field strength (4.65) has the same temperature dependence as the impulse thermal critical field (6.48). As pointed out by Vermeer (1959), it is therefore impossible to distinguish between the two mechanisms on the basis of temperature dependence of the breakdown strength; the dependence on time of application of the field is required to settle the question.

In a long series of measurements, Keller (1951) and Vermeer (1954, 1956a, b, c, 1959) showed that breakdown in various glasses was purely electrical up to a certain transition temperature and thermal thereafter. In addition, Vermeer (1954) has measured the high-field conductivity and used the results to calculate an impulse thermal critical field strength. A full account of the results for different glasses is given by Vermeer (1959); the absolute characteristics depend strongly on the sodium content, but since similar relative characteristics are observed in all cases we shall confine our comments to one of them—the borosilicate Pyrex glass BSI, for which the results of breakdown experiments will be given. The experimental arrangement used is shown in Fig. 1.3; the test voltage increased almost linearly with time, and the time to breakdown could be varied between 10^{-5} s and 30 s. Vermeer found that annealing the specimens before breakdown did not reduce the scatter of experimental results, but that cleaning in hydrofluoric acid immediately before the attachment of electrodes gave a marked reduction in scatter.

The effect of specimen thickness for various temperatures and rise times is shown in Figs. 8.16 and 8.17. The breakdown

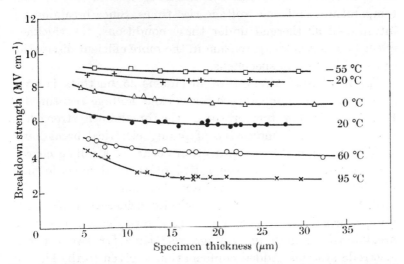

Fig. 8.16. The effect of specimen thickness on the breakdown strength of glass BSI for a voltage rise time of 30 seconds at different temperatures with liquid electrodes (after Vermeer 1954).

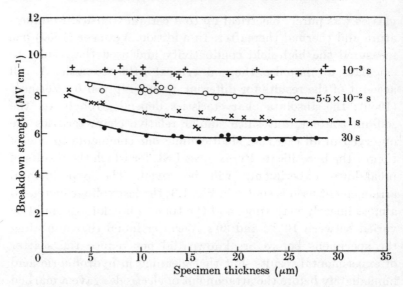

FIG. 8.17. The effect of specimen thickness on the breakdown strength of glass BSI at 20 °C for different voltage rise times with liquid electrodes (after Vermeer 1954).

strength shows appreciable thickness dependence only for high temperatures and slow voltage rise times. Since breakdown is interpreted as thermal under these conditions, the thickness effect finds a ready explanation in the more efficient dissipation of heat by the thin specimens.

The breakdown strength of specimens 25 μm thick is given as a function of temperature for various voltage rise times in Fig. 8.18. At low temperatures, the breakdown strength is constant, which is indicative of purely electrical breakdown; at higher temperatures a transition occurs to a falling characteristic, which is highly suggestive of thermal breakdown in view of the dependence on voltage rise time exhibited in Fig. 8.18. Since this dependence is a vital consideration for identifying thermal breakdown, Fig. 8.19 shows the dependence of breakdown strength on voltage rise time for two different electrode systems. Added confirmation is given to the identification of breakdown as thermal (above the transition) by the higher values obtained for liquid electrodes; these have a

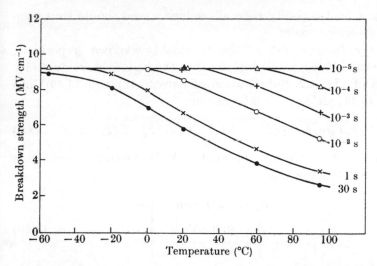

Fig. 8.18. The effect of temperature on the breakdown strength of glass BSI for different voltage rise times with liquid electrodes and specimen thickness 25 μm (after Vermeer 1954).

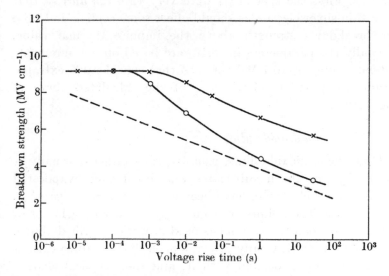

Fig. 8.19. The effect of voltage rise time on the breakdown strength of glass BSI at 20 °C. × Liquid electrodes. ○ Silver-layer electrodes. -- Theoretical results from eqn (8.9). (After Vermeer 1956b.)

greater heat capacity than silver layers, and should thus keep the specimen cooler.

As a further test of the thermal-breakdown hypothesis, Vermeer (1956b) fitted eqn (6.42) to the high-field conductivity data of Fig. 4.2. Substitution of eqns (6.42) and (6.46) into eqn (6.44) yields after integration

$$(C_V k_0 T_0^2/\sigma_0 \phi)\exp(\beta F_c + \phi/k_0 T_0) \simeq (\beta^2 F_c^2 - 2\beta F_c + 2)/\beta^3, \quad (8.9)$$

from which F_c can be computed. The conductivity parameters for BSI glass were found to be

$$\left.\begin{aligned} \sigma_0 &= 12 \ \Omega^{-1} \ \mathrm{cm}^{-1} \\ \phi &= 0{\cdot}96 \ \mathrm{eV} \\ \beta &= 2{\cdot}8 \ \mathrm{MV}^{-1} \ \mathrm{cm} \end{aligned}\right\}, \quad (8.10)$$

and (8.9) and (8.10) were used together to find the theoretical result in Fig. 8.19. The agreement with experimental breakdown results is only fair, and two principal reasons probably account for this. First the specimens were very thin (25 μm) so that even thin silver electrodes should effect some cooling and raise the breakdown strength above the impulse thermal value. Secondly the parameters in (8.10) are based on measurements extending only up to 1 MV cm^{-1}, so that considerable extrapolation is required to obtain any of the calculated curve of Fig. 8.19.

(c) Thin silicon dioxide films

Thin silicon dioxide films, usually grown either thermally or anodically on silicon substrates and having an evaporated metal counter-electrode, have been extensively investigated in recent years. These films are commonly thinner by a factor of 100 or more than the specimens used in work described above. It is found that thermally grown silicon dioxide always has a positive charge associated with it, and the electrical characteristics of the silicon–silicon-oxide interface are reviewed by Lamb (1970) and Sze (1969).

The high-field conduction properties of this MOS structure were interpreted by Lenzlinger and Snow (1969) as being due to Fowler–Nordheim emission, and their work has been discussed in § 3.1; the nominal thickness of the films used was 1000 Å. Korzo (1969) investigated the high-field conduction properties of considerably thinner films and his results are shown in Fig. 8.20. The almost horizontal part of the characteristics is taken as evidence of strong capture of injected carriers by traps, particularly at low temperatures. The steeply

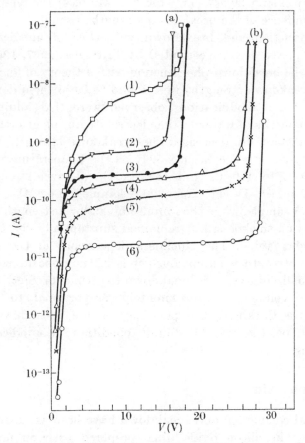

Fig. 8.20. Current–voltage characteristics of SiO$_2$ films at various temperatures (K): (1) 223; (2) 173; (3) 77; (4) 210; (5) 130; (6) 80. The film thicknesses in Å were (a) 80, (b) 300. (After Korzo 1969.)

rising current characteristic is interpreted as corresponding to the total filling of traps, and consequent strong collision ionization from them. These two events are not necessarily connected; the rapid rise in current shown can scarcely be attributed to the trap-filled limit in the sense of § 5.2(a), since eqn (5.13) predicts a trap-filled-limit voltage proportional to the square of the thickness. It is much more likely that the steep rise is associated with the onset of collision ionization of deep traps. The breakdown strength is about 10 MV cm^{-1} for 300 Å samples and 20 MV cm^{-1} for 80 Å samples; the temperature dependence of the breakdown strength is small.

The conduction and breakdown properties of anodically grown SiO$_2$ have been investigated by Fritzsche (1967, 1969), who links the breakdown phenomenon with a theory of anodic growth. Breakdown strengths were found to be of the order of 10 MV cm^{-1} and modifications observed were that addition of small quantities of water to the electrolyte during growth of the film substantially decreased the breakdown strength; the breakdown strength decreased slowly as the temperature rose from -20 °C to 100 °C, and the d.c. breakdown strength increased from 1·6 MV cm^{-1} for a 390 Å sample to 14·2 MV cm^{-1} for a 2200 Å sample, while the impulse breakdown strength did not depend on thickness, but remained substantially constant at the higher value. The thickness dependence of the d.c. breakdown strength is anomalous—it is of the opposite sense to that usually observed in breakdown experiments. Since the breakdown voltage is approximately proportional to the square of the thickness, it is possible that d.c. breakdown is consequent on the trap-filled-limit condition in anodically grown films.

8.3. Silicon oxide

Electrical conduction and breakdown have been extensively investigated in silicon oxide films prepared between metal electrode films by vacuum evaporation. The high-field conduction process has been discussed in §4.3(b), and thermal breakdown of the films in §6.2(c).

Purely electrical breakdown has been investigated by Klein and his co-workers (Klein 1966, 1969, Klein and Gafni 1966, Klein and Burstein 1969) and also by Budenstein and Hayes (1967). In all this work self-healing breakdowns were used in order to eliminate weak spots, and also to facilitate the collection of data by enabling hundreds of tests to be carried out on a single sample. No permanent short-circuit occurs with a self-healing breakdown, which causes a small pin-hole through the dielectric along the breakdown path and a larger hole in the metallic film electrode (cf. Klein 1969). The limited extent of the destruction is due to the protective effect of a series resistor. In the silicon oxide capacitors tested by Klein and co-workers (see Klein 1969) the dielectric thickness was commonly in the range 3000–5000 Å, the capacitance of order of 300 pF, and the protective resistor varying from 1 kΩ to 1 MΩ. Under these circumstances, two kinds of breakdown were observed, which were designated as single-hole breakdown and propagating breakdown.

Single-hole breakdown is associated with relatively large values of the protective resistor and small values of the applied voltage. When breakdown occurs, the voltage decreases to a minimum during the discharge and then, since the breakdown is self-healing, the silicon oxide capacitor recharges from the supply through the protective resistor. If the applied voltage is maintained, breakdowns continue to occur but at a decreasing rate. It was assumed that the destruction occurring during the breakdown was due to the discharge of the energy stored in the silicon oxide capacitor, and Klein (1966) made estimates to verify this assumption. It was found that destruction commenced in a discharge path less than 1 μm in diameter in the oxide and the metal, and that the hole in the metal electrode enlarged to some tens of μm in diameter, being larger for larger applied voltages. Klein also estimated that current densities of order 10^6 to 10^{10} A cm^{-2} occur in the discharge paths, with power input densities up to 10^9 W cm^{-2} around the edge of the vaporizing metal electrode. Analysis of the discharge spectrum showed average temperatures of the order of 4000 K for the vaporization of the electrodes.

10

If the series protective resistor was reduced or the applied voltage raised, propagating breakdown was observed. Two distinct types were observed by Klein (1966). In the first type a series of single-hole breakdowns occurs at adjacent sites. This is explained by the rapid recharge of the capacitor through the reduced series resistor; before the hot periphery of the original discharge path has cooled sufficiently a subsequent discharge occurs at a lower voltage. In the second type of propagating breakdown, an arc occurs between the pit of the initial single-hole breakdown and the metal of the counter-electrode, causing evaporation over a large area.

Full details of breakdown in silicon oxide are given by Klein (1969); his results emphasize the point that comparisons of breakdown-strength data can be validly made only if the sample fabrication is the same and if electrical power is supplied in the same way.

8.4. Polymers

The dielectric breakdown of polymers has been reviewed by Whitehead (1951) and by Mason (1959); we shall consider briefly the general features of conduction and breakdown in polymers, and also discuss in detail some theoretical interpretations that have been given for data on polyethylene and polyethylene terephthalate (mylar).

(a) Conduction processes

Low-field conduction in polymers has been interpreted as ionic or electronic depending on the polymer in question and on other conditions such as temperature, impurities, and heat treatment. Saito, Sasake, Nakajima, and Yada (1968) propose an ionic conduction mechanism for both plasticized and unplasticized polyvinyl chloride, mainly on the basis of a negative pressure dependence on the conductivity; this interpretation was extended to high-field conduction by Ieda et al. (1970) as discussed in §4.1(a). Kosaki, Ohshima, and Ieda (1970) similarly attribute the low-field conduction in polyvinyl fluoride to ions. Seanor (1967, 1968) investigated the conductivity of Nylon 66 from room temperature up to 150 °C and found protonic conduction in the upper portion of the temperature range while electronic conduction dominated in the lower part.

In addition to the work described in § 5.4(*b*) conduction in various polythenes at both low and high field strengths has been studied by many authors using various techniques of measurement. Ieda, Sawa, Morita, and Shinohara (1968) measured the temperature dependence of the low-field conductivity of low-density polythene, both pure and with metallic salts as impurities. With the data interpreted as a thermally activated process for a temperature range of 70 °C to 105 °C, values of the activation energy were found to range from 0·60 eV for pure samples up to 1·68 eV for samples with 1 per cent impurity concentration.

Electrolyte-solution electrodes were used by Swan (1966, 1967), one or both of the electrodes being prepared with iodine in solution. The steady-state current (which took many hours to establish) was found to be strongly dependent on iodine concentration; current–temperature measurements indicated an activation energy of about 1 eV for all iodine concentrations. The field-strength dependence of the current is shown in Fig. 8.21. The current is approximately proportional to the cube of the voltage; this is a somewhat more rapid increase than would be predicted by a space-charge-limited theory (cf. eqn (5.10)), but space charge may nevertheless be one of the factors contributing to the non-ohmic behaviour.

Martin and Hirsch (1969, 1970) used a time-of-flight method to measure the effective mobility of carriers that were volume-generated by electron bombardment. They distinguish between a 'prompt' and a 'long-lived' conduction process, the former being supposed to be due to the drift of excited carriers along the chain in which they are created until trapped for the first time, and the latter to subsequent drift, either in the same chain or to an adjacent one. Their results for effective mobilities of the long-lived current in polystyrene and polyethylene

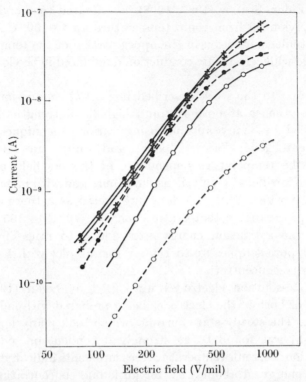

Fig. 8.21. Steady-state current against electric field strength curves for 3 mil polyethylene film at 25 °C. (1 mil = $\frac{1}{1000}$ inch.) Iodine concentrations in the electrolytic electrodes were: $+$ 100 gl^{-1}, $\bullet$ 20 gl^{-1}, $\bigcirc$ 1 gl^{-1}. The dashed lines refer to iodine in the anode, the full lines to iodine in the cathode. (After Swan 1966.)

terephthalate are shown in Fig. 8.22. It is clear that a simple activation-energy formula for the mobility (eqn (2.105)) is not valid over any extended temperature range, and that activation energies are markedly different even for temperatures only 60 °C apart. Data for the conduction parameters are collected in Table 8.13.

Davies (1972) used a surface charge dissipation technique to measure the charge-carrier mobility for various polyethenes; the relevant conduction parameters are shown in Table 8.13. Also shown are trap densities that were estimated by Davies in

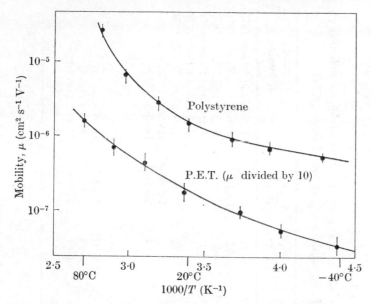

FIG. 8.22. Temperature dependence of mobility in polystyrene (holes) and polyethylene terephthalate (electrons) for a field of 1.2 MV cm^{-1} (After Martin and Hirsch (1969)).

the following way. The pre-exponential term μ_0 of eqn (2.105) is interpreted in the same way as that appearing in eqn (2.5) for ionic hopping, viz.

$$\mu_0 = a^2 e v_0/k_0 T = e v_0/k_0 T N_t^{\frac{2}{3}}, \qquad (8.11)$$

where v_0 is a characteristic frequency (taken by Davies to be 10^{14} s^{-1}), and N_t is the trap density. The trap density is not quoted for those cases in which it would yield a figure higher than the atomic density in the solid, and this fact appears to be sufficient evidence that the thermally activated hopping model is not appropriate in those cases. In the cases for which the density of trapping levels is a reasonable value, it is probably an underestimate by as much as several orders of magnitude, since eqn (8.11) does not contain the distance between trapping centres in the same way as does eqn (2.117). If the proportion

TABLE 8.13

Conduction parameters of various polymers; the experimental data are interpreted by an activation formula for the mobility (cf. eqn (2.105)), and the trap density is found from (8.11)

Substance	Temperature	μ $(\mathrm{cm^2\,V^{-1}\,s^{-1}})$	μ_0 $(\mathrm{cm^2\,V^{-1}\,s^{-1}})$	W_H (eV)	N_t $(\mathrm{cm^{-3}})$	Reference
Polystyrene	20 °C	10^{-6}	2.75×10^{-3}	0·2		Martin and Hirsch 1970
	80 °C	5×10^{-6}	2.41×10^{6}	0·75	7×10^{13}	Martin and Hirsch 1970
Polyethylene terephthalate	20 °C	1.5×10^{-6}	4.13×10^{-3}	0·2		Martin and Hirsch 1970
	80 °C	2×10^{-5}	0·375	0·3		Martin and Hirsch 1970
High-density polyethylene	80 °C	4.5×10^{-10}	3×10^{-2}	0·55		Martin and Hirsch 1970
	95 °C	9×10^{-10}				Martin and Hirsch 1970
	113 °C	1.3×10^{-9}	5.8×10^{6}	1·2	1.2×10^{13}	Davies (1972)
Iodized high-density polyethylene	113 °C	3.9×10^{-9}	3.4×10^{6}	0·84	3×10^{13}	Davies (1972)
Low-density polyethylene						
WNC18	113 °C	10^{-7}	10^{5}	0·92	5×10^{15}	Davies (1972)
R13320	113 °C	1.2×10^{-8}	1.7×10^{2}	0·78	7×10^{19}	Davies (1972)
R9706	113 °C	3.9×10^{-9}	7	0·71	9×10^{21}	Davies (1972)
Iodized WNC18	113 °C	2×10^{-5}	2.4×10^{6}	0·85	4×10^{13}	Davies (1972)

of traps occupied is small, then (2.117) leads to

$$\mu_0 = (ev_0/k_0 T N_t^{\frac{2}{3}})\exp(-2\alpha/N_t^{\frac{1}{3}}) \tag{8.12}$$

in place of (8.11). The best fit to the data will then probably occur for lower values of W_H and higher values of N_t.

Taylor and Lewis (1971) investigated conduction in polyethylene terephthalate and polyethylene without any excitation of carriers (other than that caused by the applied field), and using conventional current and voltage measuring techniques. They found the usual long-term decay of current for a steady applied voltage, but were able to overcome this difficulty by a thermal conditioning process that allowed steady values of current to be measured over a wide range of field strength and temperature. A generalization of the Schottky theory (cf. §3.3) is proposed, in which the usual term in the energy barrier due to Coulombic image force is replaced by

$$\phi(x) = -Ce/(ax)^n, \tag{8.13}$$

where C, a, and n are adjustable positive constants. Following the usual derivation of the Schottky law, we find that thermionic emission from a cathode over an energy barrier of the from (8.13) in the presence of an applied field F is given by

$$j = j_0 \exp\{-(\phi-e\beta F^{n/(n+1)})/k_0 T\}, \tag{8.14}$$

where β is a constant function of the parameters of the energy barrier of eqn (8.13). A plot of $\log j$ against $F^{n/(n+1)}$ should then be linear, and the case $n = 1$ gives the usual Schottky plot corresponding to a Coulomb-law image force. Values of $n < 1$ correspond to long-range effects such as space charge, and values of $n > 1$ to short-range effects such as a neutral trapping layer. The values of n for best fit were computed by Taylor and Lewis (1971) and their results are shown in Fig. 8.23. The zero-field limits for the activation energy yield $\phi = 2.58$ eV for polyethylene terephthalate and $\phi = 2.14$ eV for polyethylene. Since the values obtained for the parameter n are less than unity, the results were attributed to cathode emission over a barrier whose detailed shape is due to trapped space

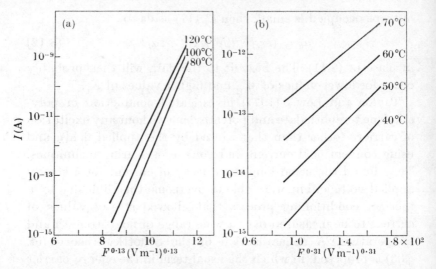

FIG. 8.23. Current–voltage characteristics plotted in the form $\log I$ against $F^{n/(n+1)}$. (a) Polyethylene terephthalate with $n = 0.15$. (b) Polyethylene with $n = 0.45$. (After Taylor and Lewis 1971.)

charge. This interpretation also provides a possible explanation for the thermal conditioning of the samples; the treatment under conditions of high field and high temperature permits the formation of the space-charge barrier in relatively short times.

Tanaka and Inuishi (1969) and Tanaka (1970) have also studied the high-field conduction of polythene, and they interpret their results as being due to a combination of Poole–Frenkel effect and space charge.

The main points made in these various experimental investigations are probably all correct, viz. that the current is due to field-dependent hopping transfer of electrons (or holes) and that the electrode emission is modified by space charge. However, the details of the interpretation may well require revision as more of the factors that lead to field dependence of the current are incorporated into the theory.

(b) Breakdown

The breakdown strength as a function of temperature is of the same general form for all polymers, that is, almost constant

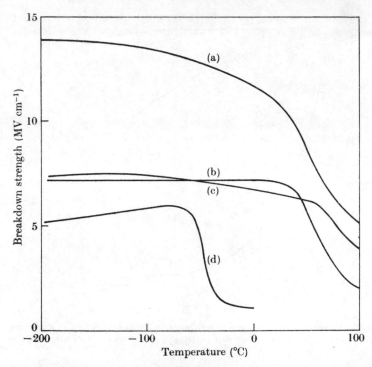

FIG. 8.24. The breakdown strength of some long-chain polymers as a function of temperature. (a) Polymethyl methacrylate. (b) Polythene. (c) Polystyrene. (d) Polyisobutylene. (After Oakes 1949.)

at low temperatures and falling off rapidly above some critical temperature. This is illustrated in Fig. 8.24 which shows results for various polymers obtained by Oakes (1949).

The similarity in the high-temperature behaviour of Fig. 8.24 is only apparent; Stark and Garton (1955) have introduced the hypothesis of electromechanical breakdown for polyethylene above 50 °C and for polyisobutylene above −40 °C. They arrive at this point of view from a consideration of the behaviour of Young's modulus with temperature, which is shown for the materials in question in Fig. 8.25. To derive the condition for electromechanical instability, consider a voltage applied to a flat-slab specimen of initial thickness d_0, causing a compressive stress $\epsilon_s V^2/8\pi d^2$, where d is the strained thickness and ϵ_s the

11

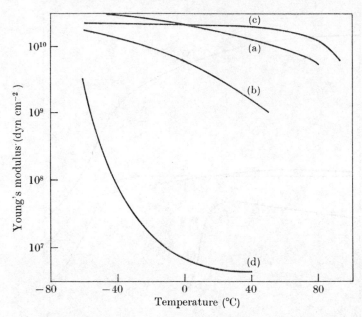

FIG. 8.25. Young's modulus (on a logarithmic scale) as a function of temperature for various polymers identified as in Fig. 8.24 (after Moll and Leferre
1948).

static dielectric constant. For large strains

$$\epsilon_s V^2 / 8\pi d^2 = Y \ln(d_0/d) \qquad (8.15)$$

where Y is the Young's modulus. The function $d^2 \ln(d_0/d)$ has a
maximum with respect to d when $\ln(d_0/d) = \frac{1}{2}$, and no stable
thickness is possible for smaller values of d. This is the electromechanical critical condition, and failure occurs due to mechanical collapse at a critical field given by

$$F_c = V_c/d = (4\pi Y/\epsilon_s)^{\frac{1}{2}}. \qquad (8.16)$$

Since breakdown fields are calculated from the original measure
thickness, the apparent critical field will be

$$F_{c(\text{app})} = V_c/d_0 = 0 \cdot 61 (4\pi Y/\epsilon_s)^{\frac{1}{2}}. \qquad (8.17)$$

Substitution of the values of Y from Fig. 8.25 shows good
numerical agreement with the high-temperature portion of the

breakdown curves of Fig. 8.24 in the cases of polyethylene and polyisobutylene.

Stark and Garton (1955) show that the high-temperature fall in the breakdown strengths of polystyrene and polymethyl methacrylate cannot be explained in this way, at least up to 80 °C. There are two possible alternative explanations; either the breakdown is thermal in nature, with field-enhanced conductivity playing an important role, or else it is purely electrical with a decreasing temperature characteristic.

The electromechanical-breakdown theory received further confirmation from experiments on a new type of sample introduced by McKeown (1965). In this new arrangement, a thin polythene film between two stainless-steel balls is encapsulated in a thermosetting epoxy resin. The temperature dependence of breakdown strength for both McKeown samples and the usual type of recessed samples was measured by Lawson (1966); his results are shown in Fig. 8.26. Clearly the steep drop-off in breakdown strength occurring above 50 °C for the recessed specimens does not occur for the rigidly held McKeown specimens. However, a new problem arises, since in the region of purely electrical breakdown the strength is some 50 per cent higher in the McKeown type of specimen. The possible reasons for this were discussed by Lawson (1966) and he identified the presence of the epoxy resin as the cause of the increase in strength for reasons which were not clear. A comparison of the experimental results shown in Fig. 8.26 with the Fröhlich amorphous critical field strength (eqn (4.65)) gives a value $\Delta V = 0.06$ eV for the spread of shallow traps below the conduction levels.

Other confirmatory evidence has also been given for electromechanical failure in polythene. Blok and LeGrand (1968) measured the optical birefringence that occurred before breakdown, using the technique of Cooper and Wallace (1956), and Fava (1965) used both optical and mechanical means to verify the hypothesis.

Various experiments have been performed to elucidate the mechanism of breakdown in the region of purely electrical

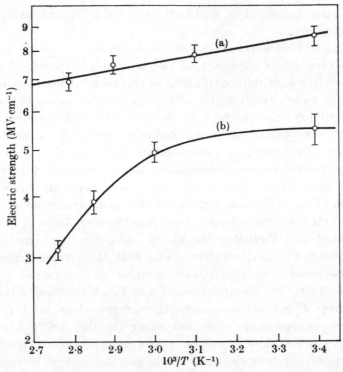

FIG. 8.26. Temperature dependence of the breakdown strength of polythene (log F^* against $10^3/T$) between 20 °C and 85 °C. (a) McKeown specimens 1965. (b) Recessed specimens. (After Lawson 1966.)

breakdown. Bradwell, Cooper, and Varlow (1971) investigated the high-field conduction and breakdown of high-purity polythene at 20 °C. In addition to d.c. and impulse tests, the effect of pre-stressing with a d.c. voltage before impulse testing was investigated, and the results of this work are shown in Fig. 8.27. When the impulse voltage has the same polarity as the pre-stressing voltage, the mean electric strength is not significantly different from that obtained with a steady d.c. voltage, but when the polarity of the impulse is opposite to that of the pre-stressing voltage a large decrease in breakdown strength occurs. The extent of the decrease is determined by the delay time, i.e. the time interval between removal of the pre-stress and application of the impulse voltage.

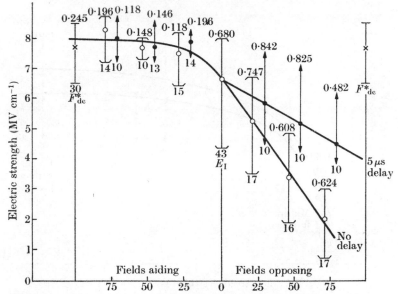

Fig. 8.27. The effect of d.c. pre-stressing on the impulse breakdown strength of high-purity polythene at 20 °C with electrodes of colloidal silver in toluene. The vertical bars represent the spread of breakdown strengths; the figure below a bar is the number of tests and the figure above is the variance. Open circles correspond to application of the impulse voltage immediately after pre-stressing, and full circles correspond to a 5 μs delay. (After Bradwell *et al.* 1971.)

The condition for 5 μs delay is shown in Fig. 8.27, and the dependence on various delay times is shown in Fig. 8.28; the electric strength has almost completely recovered after a delay of 10^{-2} s. Bradwell *et al.* (1971) also take into account observations of current transients to propose a space-charge-modified breakdown mechanism. Since the impulse breakdown strength was measured to be 86 per cent of the d.c. breakdown strength, it was assumed that the impulse test corresponded very closely to a uniform applied field, and the following assumptions were made concerning the space charge.

(1) The accumulation of space charge is negligible during the duration of an impulse test.

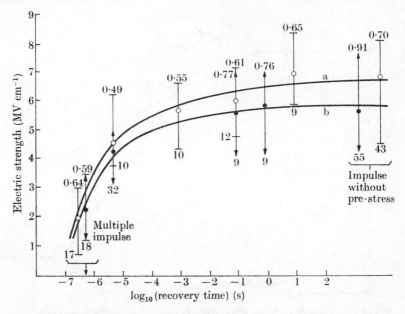

FIG. 8.28. Impulse breakdown strength of polythene after 75 per cent pre-stress of opposite polarity at 20 °C. The notation on the vertical bars is the same as in Fig. 8.27. Open circles are for high-purity material and full circles are for commercial-grade polythene. (After Bradwell *et al.* 1971.)

(2) During d.c. pre-stressing, negative space charge accumulates near the cathode and positive space charge near the anode; the density of the former exceeds that of the latter.

(3) The negative-carrier mobility exceeds that of the positive carriers, and the cathode is partially blocking to the passage of positive carriers.

This analysis accounts for the variation of field strength within the dielectric as a result of the injection of very low-mobility carriers from the electrodes, and assumes that breakdown will occur when the field strength exceeds some critical value. However, the possibility of further space-charge distortion during the breakdown process still remains; in fact all of the above data seem to be consistent with the assumption that collision ionization causes an enhanced field adjacent to the cathode, as discussed in §7.3. The recovery of breakdown strength shown in Fig. 8.28 is taken as evidence of the decay of the space charge after removal of the field. Bradwell *et al.* (1971) estimate the space-charge carrier density to be of order 3×10^{15} cm^{-3}, and the results of Fig. 8.28 imply a decay time of order 10^{-2} s. Use of these estimates in eqn (1.3a) yields a mobility of 5×10^{-8} cm^2 V^{-1} s^{-1}, which compares reasonably well with other data.

Extensive investigations of the breakdown process in polyethylene terephthalate were carried out by Riehl, Baessler, Hunklinger, Spannring, and Vaubel (1969), who developed a technique of non-destructive breakdown similar to the self-healing breakdowns observed by Klein (1969) in SiO, but with the discharge restricted before it had caused any electrical damage. The current density in the discharge was estimated at 2×10^3 A cm^{-2}, which is orders of magnitude lower than estimates for destructive discharges. The basic mechanism for initiation of a true discharge is assumed to be the same as that for a non-destructive discharge since the build-up of current is the same for both in the initial stages.

Near 4 K the pre-breakdown current is identified with Fowler–Nordheim emission (eqn (3.22)) of electrons from the cathode over a barrier of 1·8 eV, which agrees reasonably well with the values obtained by Lilly, Lowitz, and Schug (1968), quoted in §5.4(b), for thermionic emission at higher temperatures. The current densities observed in non-destructive breakdown would require the cathode field to be enhanced above the mean field by a factor of about three. This enhancement is attributed to a space charge of relatively immobile holes, which are the product of collision ionization exactly in the manner discussed in §7.3. A detailed analysis of the data has been given by Baessler, Riehl, and Spannring (1969), taking into account all features of the experimental set-up.

CONCLUSIONS

9.1. High-field conduction

THE situation regarding high-field conduction in dielectrics can best be summed up by saying that the process is not well understood. The difficulties are of various origins. In some dielectrics, the low-field conduction process is well understood, but at high fields it is supplanted by other processes, the details of which have not been determined; in other dielectrics, the resistivity is so high that low-field conduction processes are scarcely available to measurement and all observations are in the non-linear region.

In general, high-field conduction measurements have sought to identify the mechanism by measuring the dependence of current on voltage and temperature. Theories of the type discussed in Chapters 3 and 4, in which the current is either bulk-limited or electrode-limited, take current density and field strength as basic theoretical data. The current density is determined from the current simply by division by the electrode area. It is clear that the bulk of the current is carried in a very much smaller area during purely electrical breakdown, but there do not appear to be any experimental data that verify the assumption of uniform current density at high fields; proof that current is proportional to electrode area is insufficient to establish this point, since the same would be true of current carried in a large number of filamentary inter-electrode paths.

The field strength is likewise determined from the voltage by division by inter-electrode distance (in the case of plane parallel electrodes), and the assumption of a uniform field is essentially correct for either electrode or bulk-limited processes. However there do exist theories that take account of possible non-uniformity of the field (they are discussed in Chapter 5), and they are paralleled by experimental work in certain cases.

Thus the very first interpretive step, viz. that of establishing correspondence between the basic experimental and theoretical data, is at least to some extent uncertain.

However, granted the satisfactory determination of current density and field strength, there remain other difficulties in the interpretation of the results. Since the Poole–Frenkel effect (eqns (4.72) or (4.73)) is a popular means of explaining high-field conduction, the difficulties can be illustrated with reference to it.

(1) It appears that the Poole–Frenkel effect (eqn (4.72)) differs from the Richardson–Schottky effect (eqn (3.52)) experimentally, in that a plot of $\log \sigma$ against $F^{\frac{1}{2}}$ at constant temperature is linear for the former, while for the latter the linear plot is $\log j$ against $F^{\frac{1}{2}}$. However most experimental data covers only a few decades of conduction current, and it is usually impossible to decide between the two equations.

(2) In addition, both equations contain material parameters—one a pre-exponential term, and the other appearing as a multiplying factor in the exponent. The latter is usually identified as the effective dielectric constant, since it appears thus in eqns (3.52) and (4.72), while the former is frequently not explicitly identified. In extensive series of experiments, these parameters often turn out to be dependent on thickness, which clearly precludes the simple interpretation given to them by the model from which the equations were derived.

In order to find a more acceptable theoretical result for the functional dependence of the current on field strength, temperature, and thickness, complex empirical formulae such as eqn (4.74) have been fitted to the results. Alternatively, derivations of the behaviour of more detailed models such as those summarized in eqn (4.100) can be used. Whichever approach is used, several parameters are available to fit the experimental data, and, since the current is a monotonic function of the variables, a good fit will usually be achieved. The theoretical significance of a model is greatly reduced by the number of disposable parameters.

Measurement of carrier mobility by experimental techniques

that employ some form of radiation to enhance the carrier density must also be applied with caution to situations involving the dark conductivity. If the mobility in question arises from a hopping mechanism between traps, it may be very strongly influenced by the level of occupancy of the traps. Presumably an artificially high occupation level would lead to a correspondingly high effective mobility.

Since the low-field conduction process in alkali halides is well-known to be ionic, and since the conductivity is not particularly low, it would appear that a promising start could be made on the high-field conduction problem in these substances. Murgatroyd (1970b) gave a detailed survey of earlier work and attempted an experimental resolution of the problem on KCl; he did not succeed in producing an entirely definitive result, but found that electronic conduction limited by space charge was the most acceptable explanation.

9.2. Thermal breakdown

As has been mentioned previously, almost all insulators should undergo thermal breakdown at a sufficiently high temperature, since the electrical conductivity is usually an exponentially increasing function of the temperature and the thermal conductivity a slowly decreasing function of the temperature. The equation (6.1) then completely describes the process, and it is a matter of only computational complexity to find solutions for given initial and boundary conditions and functional dependences of the conductivities. The view is taken that thermal breakdown is defined as that which is governed by eqn (6.1), or, equivalently, as breakdown that can be adequately explained by extrapolation of the reversible, pre-breakdown electrical-conduction processes.

The outstanding theoretical difficulties are therefore difficulties only of computation, but this does not mean that interpretation of results is always clear-cut. In the first place, experimental data for the high-field conductivity are not always available, and, even in cases where they are, a simple agreement between

theoretical and measured results as a function of one independent variable (say the temperature) can usually not be considered sufficient to put the issue beyond doubt; this objection
is especially forceful if the breakdown data themselves are used
to estimate some parameter of the equations.

9.3. Purely electrical breakdown

Early experimental work on purely electrical breakdown was
heavily concentrated in the determination of temperature
dependence of breakdown strength and the comparison of d.c.
and impulse data. Accompanying theoretical work, particularly
that of Fröhlich on ionic crystals (cf. § 7.1), calculated an
intrinsic critical field strength, which correctly predicted both
the order of magnitude and the temperature dependence of the
breakdown strength without any disposable parameters.
Although the simple concept of intrinsic breakdown has
required modification, this impressive agreement between
theory and experiment must lead one to the conclusion that
there is a substantial element of validity in the theoretical ideas
behind the calculation of the intrinsic critical field.

Extensive experimental work by Cooper and his co-workers
(cf. § 8.1) was undertaken to elucidate the influence of various
factors such as impurity content, state of mechanical strain,
electrode material, crystallographic direction, etc., on the
breakdown strength, especially of the alkali halides. Although
this work produced much valuable breakdown data on the
effects investigated, it did not for the most part lead to hypotheses concerning the actual breakdown mechanism.

The thickness dependence of purely electrical breakdown had
long been known to be a small effect—in fact, the relative
independence of the breakdown strength on thickness was one
of the factors that led to the development of the concept of
intrinsic breakdown. However, the compilation of experimental
results shown in Fig. 7.5 established that the breakdown
strength of NaCl changed by a factor of order 10 over a range
of three decades of thickness; one cannot place too much reliance
on these data, since the thickness range is so great that different

experimental techniques had to be used at various parts of the range. However, even with this reservation, the results clearly indicate a substantial thickness dependence of breakdown strength, and the need for a similarly substantial modification of the idea of intrinsic breakdown.

What appears to be the best currently available concept of purely electrical breakdown was proposed independently by Cooper and Elliott (1966) and by O'Dwyer (1965), the former because of direct analysis of experimental data and the latter for theoretical reasons. The idea has since received further experimental support from the work of Paracchini (1971) and Riehl, Baessler, and their co-workers (Riehl *et al.* 1969, Baessler, Riehl, and Spannring 1969). The proposed steps in the breakdown process, which was described in § 7.3(*a*), are briefly as follows.

(1) Initially the field strength is uniform across the dielectric, and at a certain critical field substantial collision ionization begins within the dielectric. There does not appear to be any reason why this critical field should not be identified with an intrinsic critical field; in fact it appears logical to do so, and it should exhibit the dependence on temperature and other material parameters predicted for the appropriate intrinsic critical field.

(2) The products of collision ionization are relatively mobile electrons and relatively immobile holes. The holes drift slowly towards the cathode and their space charge distorts the field, making it stronger near the cathode and weaker near the anode.

(3) The stronger cathode field causes enhanced electron emission, which in turn results in increased collision ionization (particularly in the region adjacent to the cathode where the field is strongest) with consequent further increase in field distortion. This whole sequence of events results in a type of positive-feedback situation.

(4) The dielectric is destroyed by massive electron emission from the cathode caused by the greatly enhanced field strength in the vicinity of that electrode.

An attractive feature of this hypothesis is that it retains a substantial element of the theory of intrinsic breakdown, while at the same time incorporating recent experimental results. It should be noted, however, that it does not predict that the temperature dependence of the breakdown strength will be the same as that of the intrinsic field strength that marks the onset of large collision ionization. The reason is that breakdown depends also on the establishment of space charge by the positive holes; their mobility may be strongly temperature-dependent and thus affect their capability to form a space charge.

Further development in the understanding of purely electrical breakdown clearly depends on experiments designed to identify physical mechanisms occurring in the breakdown (such as those cited in this chapter) rather than simply to measure the dependence of breakdown strength on various parameters.

APPENDIX 1—CRYSTAL PARAMETERS

Substance	Interionic distance (Å)	Reduced effective ionic mass $(1/M^{+}+1/M^{-})^{-1}$ $(10^{-24}$ gm)	Dielectric constants	
			Static ϵ_s	High-frequency ϵ_∞
LiF	2·01	8·45	9·27	1·92
LiCl	2·49	9·64	10·62	2·75
LiBr	2·64	10·6	10·95	3·10
LiI	2·87	10·9	8·29	3·80
NaF	2·31	17·4	6·00	1·74
NaCl	2·81	23·3	5·62	2·25
NaBr	2·98	29·8	6·10	2·62
NaI	3·23	32·5	6·60	2·91
KF	2·67	21·3	6·05	1·85
KCl	3·14	31·0	4·68	2·13
KBr	3·29	43·8	4·51	2·33
KI	3·53	49·9	3·94	2·69
RbF	2·80	25·9	5·90	1·93
RbCl	3·27	41·8	5·00	2·19
RbBr	3·43	68·7	5·00	2·33
RbI	3·66	84·8	5·00	2·63

FOR VARIOUS ALKALI HALIDES

Dielectric constants effective $(1/\epsilon_\infty - 1/\epsilon_s)^{-1}$	Frequency of transverse optical mode $(10^{13}\ \text{s}^{-1})$	Fröhlich polaron coupling constant $(m^* = m)$	Ionization energy (eV)	Thermal conductivity κT at room temperature (W cm^{-1})
2·42	3·90	6·4		50
3·71	3·25	4·8	8·6	
4·44	4·50	3·5	6·7	
7·01	5·00	2·4	5·6	
2·45	5·25	5·9		29
3·75	3·63	5·0	7·7	27
4·59	2·55	5·0	6·5	6·6
5·20	2·20	4·8	5·4	
2·66	4·75	5·8		19
3·91	2·71	5·8	7·6	28
4·82	2·27	5·4	6·6	10
8·48	1·91	4·8	5·6	8·2
2·87	4·35	5·7		
3·90	2·24	6·3	7·4	5·4
4·36	1·69	6·6	6·4	10·3
5·55	1·42	5·8	5·5	8·8

BIBLIOGRAPHY

ABRAMOWITZ, M. and STEGUN, I. A. (eds) (1964). *Handbook of Mathematical functions*. N.B.S. Publication.

ADAMS, A. R. and SPEAR, W. E. (1964). *J. Phys. Chem. Solids* **25**, 1113.

ADLER, D. (1968). Insulating and metallic states in transition metal oxides. *Solid St. Phys.* **21**, 1.

AHRENKIEL, R. K. and BROWN, F. C. (1964). *Phys. Rev.* **136**, A 223.

ALGER, R. S. and VON HIPPEL, A. (1949). *Phys. Rev.* **76**, 127.

ALLCOCK, G. R. (1956). *Adv. Phys.* **5**, 412.

ANDERSON, P. W. (1958). *Phys. Rev.* **109**, 1492.

ANDREEV, G. A. (1958). *Izv. Akad. Nauk. U.S.S.R.* **22**, 415.

ANTULA, J. (1971). *J. appl. Phys.* **42**, 2081.

APPEL, J. (1968). Polarons. *Solid St. phys.* **21**, 193.

AUSTEN, E. W. and WHITEHEAD, S. (1940). *Proc. R. Soc.* A **176**, 33.

AUSTIN, J. G. and MOTT, N. F. (1969). *Adv. Phys.* **18**, 41.

BAESSLER, H., RIEHL, N., and SPANNRING, W. (1969). *Z. für angew. Phys.* **27**, 321.

BARAFF, G. A. (1962). *Phys. Rev.* **128**, 2507.

BARTON, J. L. (1970). *J. non-cryst. Solids* **4**, 220.

BASSANI, F. and FUMI, F. G. (1954). *Nuovo Cim.* **11**, 274.

BAUER, C. F. and WHITMORE, D. H. (1970). *Phys. Stat. Solidi* **37**, 585.

BEAUMONT, J. H. and JACOBS, P. W. M. (1966). *J. Chem. Phys.* **45**, 1496.

BLATT, F. J. (1968). *Physics of electronic conduction in solids*. McGraw-Hill.

BLOK, J. and LEGRAND, D. G. (1968). *Report of Conference on Electric Insulation and Dielectric Phenomena*, p. 53.

BORN, M. and HUANG, K. (1954). *Dynamical theory of crystal lattices*. Oxford.

BOSMAN, A. J. and VAN DAAL, H. J. (1970). *Adv. Phys.* **19**, 1.

BRADWELL, A. and PULFREY, D. L. (1968). *J. Phys.* D, **1**, 1581.

—— COOPER, R., and VARLOW, B. (1971). *Proc. I.E.E.* **118**, 247.

BROWN, F. C. and INCHAUSPÉ, N. (1961). *Phys. Rev.* **121**, 1303.

BUDENSTEIN, P. P. and HAYES, P. J. (1967). *J. appl. Phys.* **38**, 2837.

—— HAYES, P. J., SMITH, J. L., and SMITH, W. B. (1969). *J. Vacuum Sci. Tech.* **6**, 289.

BURGESS, R. E., KROEMER, H., and HOUSTON, J. M. (1953). *Phys. Rev.* **90**, 515.

CALDERWOOD, J. H. and COOPER, R. (1953). *Proc. phys. Soc., Lond.* B **66**, 73.

—— —— and WALLACE, A. A. (1953). *Proc. I.E.E.* **100**, IIA, 105.

CALLEN, H. B. (1949). *Phys. Rev.* **76**, 1394.

—— and OFFENBACHER, E. L. (1953). *Phys. Rev.* **90**, 401.

CASPARI, M. E. (1955). *Phys. Rev.* **78**, 1679.

CASTNER, T. G. and KÄNZIG, W. (1957). *J. Phys. Chem. Solids* **3**, 178.

CHOU, S. and BROOKS, H. (1970). *J. appl. Phys.* **41**, 4451.

COHEN, M. H. (1971). *Physics to-day.* **24**, no. 5, 26.

CONWELL, E. M. (1967). High field transport in semiconductors. *Solid St. Phys.* Supp. 9.

COPPLE, C., HARTREE, D. R., PORTER, A., and TYSON, H. (1939). *Proc. Inst. elect. Engrs.* **85**, 56.

COOPER, R. (1962). The electric breakdown of alkali halide crystals, *Prog. Dielect.* **5**, 97.

—— and ELLIOTT, C. T. (1966). *Br. J. appl. Phys.* **17**, 481.

—— —— (1968). *J. Phys.* D **1**, 121.

—— and FERNANDEZ, A. (1958). *Proc. phys. Soc., Lond.* B **71**, 688.

—— and GROSSART, D. T. (1953). *Proc. phys. Soc., Lond.* B **66**, 716.

—— and PULFREY, D. L. (1971). *J. Phys.* D **4**, 292.

—— and SMITH, W. A. (1961). *Proc. phys. Soc., Lond.* B **78**, 734.

—— and WALLACE, A. A. (1953). *Proc. phys. Soc., Lond.* B **66**, 1113.

—— —— (1956). *Proc. phys. Soc., Lond.* B **69**, 1287.

—— GROSSART, D. T., and WALLACE, A. A. (1957). *Proc. phys. Soc., Lond.* B **70**, 169.

—— HIGGIN, R. M., and SMITH, W. A. (1960). *Proc. phys. Soc., Lond.* B **76**, 817.

DAVIES, D. K. (1972). *J. Phys.* D **5**, 162.

DAVISSON, J. W. (1946). *Phys. Rev.* **70**, 685.

—— (1959). Directional breakdown effects in crystals, *Prog. Dielect.* **1**, 61.

DIGNAM, M. J. (1968). *J. Phys. Chem. Solids* **29**, 249.

DUKE, C. B. (1969). Tunnelling in solids. *Solid St. Phys.* Supp. 10.

EMTAGE, P. R. (1967). *J. appl. Phys.* **38**, 1820.

—— (1971). *Phys. Rev.* B **3**, 2686.

—— and O'DWYER, J. J. (1966). *Phys. Rev. Lett.* **16**, 356.

ETZEL, H. W. and MAURER, R. J. (1950). *J. Chem. Phys.* **18**, 1003.

FAVA, R. A. (1965). *Proc. Inst. elect. Engrs.* **112**, 819.

FEYNMAN, R. P. (1955). *Phys. Rev.* **97**, 660.

FEYNMAN, R. P., HELLWARTH, R. W., IDDINGS, C. K., and PLATZMAN, P. M. (1962). *Phys. Rev.* **127**, 1004.

FISHER, J. C. and GIAEVER, I. (1961). *J. appl. Phys.* **32**, 172.

FOCK, V. (1927). *Arch. Elektrotech.* **19**, 71.

FORLANI, F. and MINNAJA, N. (1964). *Phys. Stat. Solidi* **4**, 311.

—— —— (1969). *J. Vacuum Sci. Tech.* **6**, 518.

FOWLER, R. H. and NORDHEIM, L. (1928). *Proc. R. Soc.* A **119**, 173.

FRANZ, W. (1939). *Z. Phys.* **113**, 607.

—— (1952). *Z. Phys.* **132**, 285.

—— (1956). 'Dielektrischer Durchschlg', *Handbuch der Physik*, vol. 17. Springer-Verlag.

FRENKEL, J. (1930). *Phys. Rev.* **36**, 1604.

—— (1938). *Phys. Rev.* **54**, 647.

FRITZSCHE, C. R. (1967). *Z. angew. Phys.* **24**, 48.

—— (1969). *J. Phys. chem. Solids* **30**, 1885.

FRÖHLICH, H. (1937). *Proc. R. Soc.* A **160**, 230.

—— (1940). *E.R.A. Rep. L/T.* 113 (1940).

—— (1947a). *Proc. R. Soc.* A **188**, 521.

—— (1947b). *Proc. R. Soc.* A **188**, 532.
—— (1952). *E.R.A. Report L/T.* 277.
—— (1954). *Adv. Phys.* **3**, 325.
—— and PARANJAPE, B. V. (1956). *Proc. phys. Soc., Lond.* B **69**, 21.
—— and SEITZ, F. (1950). *Phys. Rev.* **79**, 526.
—— and SEWELL, G. L. (1959). *Proc. phys. Soc., Lond.* **74**, 643.
——, PELZER, H. and ZIENAU, S. (1950). *Phil. Mag.* **41**, 221.
FULLER, R. G. and REILLY, M. H. (1967). *Phys. Rev. Lett.* **19**, 113.
——, REILLY, M. H., MARQUARDT, C. L., and WELLS, J. C. (1968). *Phys. Rev. Lett.* **20**, 663.
GADZUK, J. W. (1970a). *J. appl. Phys.* **41**, 286.
—— (1970b). *Phys. Rev.* B **1**, 2110.
GELLER, M. (1956). *Phys. Rev.* **101**, 1685.
GIBBONS, D. J. and SPEAR, W. E. (1966). *J. Phys. Chem. Solids* **27**, 1917.
GOOD, R. H. and MÜLLER, W. (1956). 'Field Emission', *Handbuch der Physik*, vol. 21. Springer Verlag.
GOODMAN, A. M. (1967). *Phys. Rev.* **164** 1145.
—— and O'NEILL, J. J. (1966). *J. appl. Phys.* **37**, 3580.
GRÜNDIG, H. (1960). *Z. Phys.* **158**, 577.
GUNDLACH, K. H. (1969). *Solid-St. Electron.* **12**, 13.
GURARI, M. (1953). *Phil. Mag.* **44**, 329.
HANSCOMB, J. R. (1962). *Aust. J. Phys.* **15**, 504.
—— (1969). *J. Phys.* D **2**, 1327.
—— (1970). *J. appl. Phys.* **41**, 3597.
——, KAO, K. C., CALDERWOOD, J. H., O'DWYER, J. J., and EMTAGE, P. R. (1966). *Proc. phys. Soc., Lond.* **88**, 425.
HARRIS, L. B. (1968). *Appl. Phys. Lett.* **13**, 154.
HARTKE, J. L. (1968). *J. appl. Phys.* **39**, 4871.
HARTMAN, T. E. and CHIVIAN, J. S. (1964). *Phys. Rev.* **134**, A1094.
——, BLAIR, J. C., and BAUER, R. (1966). *J. appl. Phys.* **37**, 2468.
HEIKES, R. R. and JOHNSTON, W. D. (1957). *J. Chem. Phys.* **26**, 582.
HELLER, W. R. (1951). *Phys. Rev.* **84**, 1130.
HEYES, W. and WATSON, D. B. (1968). *Nature, Lond.* **220**, 572.
HILL, R. M. (1971). *Phil. Mag.* **23**, 59.
VON HIPPEL, A. (1931). *Z. Phys.* **67**, 707.
—— (1934). *Z. Phys.* **88**, 358.
—— (1935a). *Ergebn. exakt. Naturw.* **14**, 79.
—— (1935b). *Ergebn. exakt. Naturw.* **14**, 79.
—— and LEE, G. M. (1941). *Phys. Rev.* **59**, 824.
—— and MAURER, R. J. (1941). *Phys. Rev.* **59**, 820.
——, GROSS, E. P., JELATIS, J. G., and GELLER, M. (1953). *Phys. Rev.* **91**, 568.
HIRTH, H. and TÖDHEIDE-HAUPT, U. (1969). *Phys. Stat. Solidi* **31**, 425.
HOLM, R. (1951). *J. appl. Phys.* **22**, 569.
HOLSTEIN, T. (1959). *Ann. Phys.* **8**, 343.
HOWARTH, D. J. and SONDHEIMER E. H. (1953). *Proc. R. Soc.* A **219**, 53.
IEDA, M., KOSAKI, M., and SUGIYAMA, K. (1970). *Annual Report of the Conference on Electrical Insulation and Dielectric Phenomena*, p. 17.

——, SAWA, G. and KATO, S. (1971). *J. appl. Phys.* **42**, 3737.

—— ——, MORITA, S. and SHINOHARA, U. (1968). *J.I.E.E. Japan* **88**, 22.

INGE, L. and WALTHER, A. (1930). *Z. Phys.* **64**, 830.

——, SEMENOFF, N., and WALTHER, A. (1925). *Z. Phys.* **32**, 273.

JAIN, S. C. and PARASHAR, D. C. (1969). *J. Phys.* C **2**, (2), 167.

JACOBS, P. W. M. and TOMPKINS, F. C. (1952). *Q. Rev.* **4**,

JONSCHER, A. K. (1967). *Thin Solid Films* **1**, 213.

—— and ANSARI, A. A. (1971). *Phil. Mag.* **23**, 205.

KAWAMURA, H., OHKURA, H., and KIKUCHI, T. (1954). *J. phys. Soc. Japan* **9**, 541.

KELLER, F. J., MURRAY, R. B., ABRAHAM, M. M., and WEEKS, R. A. (1967). *Phys. Rev.* **154**, 812.

KELLER, K. J. (1951). *Physica, 's Grav.* **17**, 511.

KELTING, H. and WITT, H. (1949). *Z. Phys.* **126**, 697.

KENNEDY, T. N. and MACKENZIE, J. D. (1967). *Phys. Chem. Classes* **8**, 169.

KLEIN, N. (1966). *IEEE Trans. electron devices.* ED**13**. 788.

—— (1969). *Adv. Electron. Electron Phys.* **26**, 309.

—— and BURSTEIN, E. (1969). *J. appl. Phys.* **40**, 2728.

—— and GAFNI, H. (1966). *IEEE Trans. electron devices.* ED **13**, 281.

—— and LISAK, Z. (1966). *Proc. Instn. Elect. Electron. Engrs.* **54**, 979.

KLINGER, M. I. (1968). *Rep. Progr. Phys.* **31**, 225.

KONOROVA, E. A. and SOROKINA, L. A. (1957). *J. exp. theor. Phys. USSR* **32**, 143.

—— —— (1965). *Fizika tverd. Tela.* **7**, 1475.

KORZO, V. F. (1969). *Sov. Phys. Sol. State.* **11**, 328.

KOSAKI, M., OHSHIMA, H., and IEDA, M. (1970). *J. phys. Soc. Japan* **29**, 1012.

KUCHIN, V. D. (1957). *Rep. Akad. Sci.* U.S.S.R. **114**, 301.

LAMB, D. R. (1967). *Electrical conduction mechanisms in thin insulating films.* Methuen.

—— (1970). *Thin Solid Films* **5**, 247.

LAMPERT, M. A. (1956). *Phys. Rev.* **103**, 1648.

—— (1964). *Rep. Progr. Phys.* **27**, 329.

—— and MARK, P. (1970). *Current injection in Solids.* Academic Press.

LANDAU, L. D. (1933). *Phys. Z. Sowjun.* **3**, 644.

LANDSBERG, P. T. (1951). *Proc. R. Soc.* A **206**, 463.

LASS, J. (1931). *Z. Phys.* **69**, 313.

LAWSON, W. G. (1966). *Proc. Instn. Elect. Engrs.* **113**, 197.

LECOMBER, P. G. and SPEAR, W. E. (1970). *Phys. Rev. Lett.* **25**, 509.

LECOMTE, J. (1958). *Handb. Phys.* **20**, 244.

LEE, T. D., LOW, F., and PINES, D. (1953). *Phys. Rev.* **90**, 297.

LENZLINGER, M. and SNOW, E. H. (1969). *J. appl. Phys.* **40**, 278.

LIDDIARD, A. B. (1957). 'Ionic conductivity', *Handb. Phys.* **20**.

LILLY, A. C. and McDOWELL, J. R. (1968). *J. appl. Phys.* **39**, 141.

——, LOWITZ, D. A., and SCHUG, J. C. (1968). *J. appl. Phys.* **39**, 4360.

LOMER, P. D. (1950). *Proc. phys. Soc., Lond.* **63B**, 818.

McCOLL, M. and MEAD, C. A. (1965). *Trans. metal. Soc. A.I.M.E.* **233**, 502.

McKeown, J. J. (1965). *Proc. Instn Elect. Engrs.* **112**, 824.

Mark, P. and Hartman, T. E. (1968). *J. appl. Phys.* **39**, 2163.

Martin, E. H. and Hirsch, J. (1969). *Solid State Communs* **7**, 783.

—— —— (1970). *J. non-cryst. Solids* **4**, 133.

Maserjian, J. (1967). *J. Phys. Chem. Solids* **28**, 1957.

—— and Mead, C. A. (1967). *J. Phys. Chem. Solids* **28**, 1971.

Mason, J. H. (1959). 'Dielectric breakdown of solid insulation', *Prog. Dielect.* **1**, 1.

Merrill, R. C. and West, R. A. (1963). *Abstracts of Spring Meeting of Electrochemical Society*, Pittsburgh.

Meyerhofer, D. and Ochs, S. A. (1963). *J. appl. Phys.* **34**, 2535.

Miller, A. and Abrahams, E. (1960). *Phys. Rev.* **120**, 745.

Miller, L. S., Howe, S., and Spear, W. E. (1968). *Phys. Rev.* **166**, 871.

Moll, H. W. and LeFevre, W. S. (1948). *Ind. Engng. Chem.* **40**, 2172.

Moon, P. H. (1931). *Trans. Am. Inst. Elect. Engrs.* **50**, 1008.

Mort, J. (1971). *Phys. Rev.* B **3**, 3576.

Mott, N. F. (1967). *Adv. Phys.* **16**, 49.

—— and Gurney, R. W. (1948). *Electronic processes in ionic crystals.* Oxford.

—— and Sneddon, I. N. (1948). *Wave mechanics and its applications.* Oxford.

—— and Twose, W. D. (1961). *Adv. Phys.* **10**, 107.

Murgatroyd, P. N. (1970a). *J. Phys.* D **3**, 151.

Murgatroyd, P. N. (1970b). Ph.D. Thesis, Bristol.

Murphy, E. L. and Good, R. H. (1956). *Phys. Rev.* **102**, 1464.

Nester, H. H. and Kingery, W. D. (1963). *International Conference on Glasses, Brussels*, p. 106.

Nettel, S. J. (1963). *Polarons and excitons* (ed. C. G. Kuper and G. D. Whitefield), p. 245. Oliver and Boyd.

Neubert, T. J. and Reffner, J. A. (1962). *J. Chem. Phys.* **36**, 2780.

Oakes, W. G. (1949). *Proc. Instn. Elect. Engrs.* **96** I, 37.

O'Dwyer, J. J. (1954). *Aust. J. Phys.* **7**, 36.

—— (1957). *Proc. phys. Soc., Lond.* B **70**, 761.

—— (1960). *Aust. J. Phys.* **13**, 270.

—— (1965). *Report of Conference on Electrical Insulation and Dielectric Phenomena*, p. 33.

—— (1966). *J. appl. Phys.* **37**, 599.

—— (1967). *J. Phys. Chem. Solids* **28**, 1137.

—— (1968). *J. appl. Phys.* **39**, 4356.

—— (1969a). *J. appl. Phys.* **40**, 3887.

—— (1969b). *J. electrochem. Soc.* **116**, 239.

—— (1969c). *Report of Conference on Electrical Insulation and Dielectric Phenomena*, p. 137.

Panofsky, W. and Phillips, M. (1962). *Classical electricity and magnetism*, 2nd edn, p. 123. Addison-Wesley.

Paracchini, C. (1971). *Phys. Rev.* B **4**, 2342.

—— and Schianchi, G. (1970). *Solid State Communs.* **8**, 1769.

Paranjape, B. V. (1961). *Proc. phys. Soc., Lond.* **78**, 516.

Parmenter, R. H. and Ruppel, W. (1959). *J. appl. Phys.* **30**, 1548.

PENLEY, J. C. (1962). *Phys. Rev.* **128**, 596.

PINES, D. (1953). *Phys. Rev.* **92**, 626.

PLESSNER, K. W. (1948). *Proc. phys. Soc., Lond.* **60**, 243.

PLUMMER, E. W. and YOUNG, R. D. (1970). *Phys. Rev.* B **1**, 2088.

POLLACK, S. R. (1963). *J. appl. Phys.* **34**, 877.

PRUETT, H. D. and BROIDA, H. P. (1967). *Phys. Rev.* **164**, 1138.

PULFREY, D. L., SHOUSHA, A. H. M., and YOUNG, L. (1970). *J. appl. Phys.* **41**, 2838.

REDFIELD, A. G. (1954). *Phys. Rev.* **94**, 526.

RIDLEY, B. K. (1963). *Proc. phys. Soc., Lond.* **82**, 954.

RIEHL, N., BAESSLER, H., HUNKLINGER, S., SPANNRING, W., and VAUBEL, G. (1969). *Z. für angew. Phys.* **27**, 261.

ROLFE, J. (1964). *Can. J. Phys.* **42**, 2195.

ROSE, A. (1955). *Phys. Rev.* **97**, 1538.

SAITO, S., SASABE, H., NAKAJIMA, T., and YADA, K. (1968). *J. Polymer Sci.* A-2, **6**, 1297.

SCHISSLER, L. R. (1960). M.I.T. Tech. Report, **153**.

SCHMID, A. P. (1968). *J. appl. Phys.* **39**, 3140.

SEAGER, C. H. and EMIN, D. (1970). *Phys. Rev.* B **2**, 3421.

SEANOR, D. A. (1967). *J. Polymer Sci.* C **17**, 195.

—— (1968). *J. Polymer Sci.* A-2, **6**, 463.

SEITZ, F. (1940). *Modern theory of solids.* McGraw-Hill.

—— (1948). *Phys. Rev.* **73**, 549.

—— (1949). *Phys. Rev.* **76**, 1376.

SEKI, H. (1970). *Phys. Rev.* B **2**, 4877.

SERVINI, A. and JONSCHER, A. K. (1969). *Thin Solid Films* **3**, 341.

SHOCKLEY, W. (1951). *Bell Syst. tech. J.* **30**, 990.

—— (1961). *Solid St. Electron.* **2**, 35.

SIMMONS, J. G. (1963a). *J. appl. Phys.* **34**, 1793.

—— (1963b). *J. appl. Phys.* **34**, 2581.

—— (1964). *J. appl. Phys.* **35**, 2472.

—— (1967). *Phys. Rev.* **155**, 657.

SMITH, J. L. and BUDENSTEIN, P. P. (1969). *J. appl. Phys.* **40**, 3491.

SOMMERFELD, A. and BETHE, H. (1933). *Handb. Phys.* **24/2**.

SPEAR, W. E. (1961). *J. phys. chem. Solids* **21**, 110.

—— and MORT, J. (1963). *Proc. phys. Soc., Lond.* **81**, 130.

STARK, K. H. and GARTON, C. G. (1955). *Nature, Lond.* **176**, 1225.

STASIW, O. and TELTOW, J. (1947). *Annln Phys.* **1**, 261.

STRATTON, R. (1957). *Proc. R. Soc.* A, **242** 355.

—— (1958). *Proc. R. Soc.* A **246**, 406.

—— (1961). 'The theory of dielectric breakdown in solids', *Progr. Dielectrics* **3**, 235.

—— (1962). *J. Phys. Chem. Solids* **23**, 1177.

SWAN, D. W. (1966). *Br. J. appl. Phys.* **17**, 1365.

—— (1967). *Annual Report of Conference on Electrical Insulation and Dielectric Phenomena*, p. 27

SZE, S. M. (1967). *J. appl. Phys.* **38**, 2951.

—— (1969). *Physics of semiconductor devices.* John Wiley.

TANAKA, T. (1970). *Report of Conference on Electrical Insulation and Dielectric Phenomena*, p. 23.
—— and INUISHI, Y. (1969). *Elec. Eng. Japan* **89**, 1.
TAYLOR, D. M. and LEWIS, T. J. (1971). *J. Phys.* D 4, 1346.
THORNBER, K. K. and FEYBMAN, R. P. (1970). *Phys. Rev.* B 1, 4099.
TOMURA, M. and KIKUCHI, T. (1952). *J. phys. Soc. Japan* **7**, 538.
UNGER, S. and TEEGARDEN, K. (1967). *Phys. Rev. Lett.* **19**, 1229.
VEELKEN, R. (1955). *Z. Phys.* **142**, 476.
VERMEER, J. (1954). *Physica, 's Grav.* **20**, 313.
—— (1956a). *Physica, 's Grav.* **22**, 99.
—— (1956b). *Physica, 's Grav.* **22**, 111.
—— (1956c). *Physica, 's Grav.* **22**, 91.
—— (1959). Thesis, Delft.
VOROB'EV, A. A. (1956). *Soviet Phys. tech. Phys.* **26**, 330.
——, VOROB'EV, G. A., and MURASHKO, L. T. (1963). *Soviet Phys. Solid State* **4**, 1441.
VOROB'EV, G. A. and PIKALOVA, I. S. (1969). *Soviet Phys. Solid State* **11**, 448.
WATSON, D. B. and HEYES, W. (1970). *J. Phys. chem. Solids* **31**, 2531.
—— ——, KAO, K. C., and CALDERWOOD, J. H. (1965). *I.E.E.E. Trans. Elec. Insul.* E1-1, 30.
WEAVER, C. and MACLEOD, J. E. S. (1965). *Br. J. appl. Phys.* **16**, 441.
WHITEHEAD, S. (1951). *Dielectric breakdown of solids.* Oxford.
—— and NETHERCOT, W. (1935). *Proc. phys. Soc., Lond.* **47**, 974.
WIJSMAN, R. A. (1949). *Phys. Rev.* **75**, 833.
WILLIAMS, R. (1964). *J. Phys. Chem. Solids* **25**, 853.
—— (1965). *Phys. Rev.* **140**, A569.
WIT, H. J. DE (1968). *Philips Res. Rep.* **23**, 449.
YABLONOVITCH, E. (1971). *Appl. Phys. Lett.* **19**, 495.
YAMASHITA, J. and KUROSAWA, T. (1958). *J. Phys. Chem. Solids* **5**, 34.
—— and WATANABE, M. (1954). *Progr. theor. Phys.* **12**, 443.
YEARGAN, J. R. and TAYLOR, H. L. (1968). *J. appl. Phys.* **39**, 5600.
ZELM, M. (1968). *Z. Phys.* **212**, 280.
ZENER, C. (1934). *Proc. R. Soc.* A 145, 523.
ZIMAN, J. M. (1964). *Principles of the theory of solids.* Cambridge University Press.

AUTHOR INDEX

SUBJECT INDEX